# TL 9000: A Guide to Measuring Excellence in Telecommunications

Also Available from ASQ:

*The TL 9000 Guide for Auditors*
Mark Kempf

*ISO 9001:2000 Explained,* Second Edition
Charles A. Cianfrani, Joseph J. Tsiakals, and John E. (Jack) West

*ISO Lesson Guide 2000: Pocket Guide to Q9001:2000,* Second Edition
Dennis Arter and J.P. Russell

*TL 9000 Quality System Requirements*
QuEST Forum

*TL 9000 Quality System Metrics*
QuEST Forum

*The Quality Audit Handbook,* Second Edition
J.P. Russell, Editing Director

*Managing Change: Practical Strategies for Competitive Advantage*
Kari Tuominen

*Root Cause Analysis: Simplified Tools and Techniques*
by Bjørn Anderson and Tom Fagerhaug

*After the Quality Audit: Closing the Loop on the Audit Process,*
Second Edition
J.P. Russell and Terry Regel


To request a complimentary catalog of ASQ Quality Press publications,
call 800-248-1946, or visit our online bookstore at http://qualitypress.asq.org .

# TL 9000: A Guide to Measuring Excellence in Telecommunications

Sandford Liebesman, Alka Jarvis,
Ashok V. Dandekar

ASQ Quality Press
Milwaukee, Wisconsin

*TL 9000: A Guide to Measuring Excellence In Telecommunications*
Sandford Liebesman, Alka Jarvis, Ashok V. Dandekar

**Library of Congress Cataloging-in-Publication Data**

Jarvis, Alka, 1947–
   TL 9000 : a guide to measuring excellence in telecommunications / Alka Jarvis, Sandford Liebesman, Ashok V. Dandekar.
      p. cm.
   Includes bibliographical references and index.
   ISBN 0-87389-509-6 (alk. paper)
   1. Telecommunications—Quality control. 2. TL 9000 (Standard) I. Title: TL9000.
II. Liebesman, Sandford, 1935– III. Dandekar, Ashok V., 1951– IV. Title.

TK5102.84 .J37 2001
384'.068—dc21                                                              2001016062

© 2001 by ASQ

All rights reserved. No part of this book may be reproduced in any form or by any means, electronic, mechanical, photocopying, recording, or otherwise, without the prior written permission of the publisher.

10 9 8 7 6 5 4 3 2 1

ISBN 0-87389-509-6

Acquisitions Editor: Ken Zielske
Project Editor: Annemieke Koudstaal
Production Administrator: Gretchen Trautman
Special Marketing Representative: David Luth

ASQ Mission: The American Society for Quality advances individual and organizational performance excellence worldwide by providing opportunities for learning, quality improvement, and knowledge exchange.

Attention: Bookstores, Wholesalers, Schools and Corporations: ASQ Quality Press books, videotapes, audiotapes, and software are available at quantity discounts with bulk purchases for business, educational, or instructional use. For information, please contact ASQ Quality Press at 800-248-1946, or write to ASQ Quality Press, P.O. Box 3005, Milwaukee, WI 53201-3005.

To place orders or to request a free copy of the ASQ Quality Press Publications Catalog, including ASQ membership information, call 800-248-1946. Visit our web site at www.asq.org or http://qualitypress.asq.org .

Printed in the United States of America

 Printed on acid-free paper

Quality Press
P.O. Box 3005
Milwaukee, Wisconsin 53201-3005
800-248-1946
Fax 414-272-1734
www.asq.org
http://qualitypress.asq.org
http://standardsgroup.asq.org

# Dedication

We dedicate this book to Henry Malec, a colleague and good friend. Henry joined the QuEST Forum at the start and contributed to the development of the Requirements and Metrics Handbooks. He was the principal creator of the Metrics Repository System, in which he used a "double blind process" to assure the integrity of the data. We relied on Henry to give us a clear view of where he thought the QuEST Forum should be heading. Sadly, he was taken from us much too early.

# Table of Contents

**Foreword** . . . . . . . . . . . . . . . . . . . . . . . . . . . . xii
**Preface** . . . . . . . . . . . . . . . . . . . . . . . . . . . . . xiii
**Acknowledgments** . . . . . . . . . . . . . . . . . . . . . . . xv
**Chapter 1 Introduction** . . . . . . . . . . . . . . . . . . . . . 1
   Creation of the QuEST Forum . . . . . . . . . . . . . . . . 1
   Telecommunications Industry Needs . . . . . . . . . . . . . 2
   The Goals and Expectations of TL 9000 . . . . . . . . . . . 3
      QuEST Forum Membership Categories . . . . . . . . . 4
      The TL 9000 Model . . . . . . . . . . . . . . . . . 5
      The TL 9000 Handbooks . . . . . . . . . . . . . . . 6
   Sources Used to Develop the TL 9000
      Requirements Handbook . . . . . . . . . . . . . . . 6
   Key Additions to ANSI/ISO/ASQC Q9001–1994 . . . . . . . 8
   The Use of Metrics to Improve Product and
      Service Quality . . . . . . . . . . . . . . . . . . . . 8
      The Metrics Implementation Process . . . . . . . . . . 9
      Approved Metrics . . . . . . . . . . . . . . . . . . 10
   Product and Service Categories . . . . . . . . . . . . . . . 11
   Pilot Program . . . . . . . . . . . . . . . . . . . . . . . 12
   Registration Process . . . . . . . . . . . . . . . . . . . . 13
   Training . . . . . . . . . . . . . . . . . . . . . . . . . . 13
   Administration . . . . . . . . . . . . . . . . . . . . . . 14
      Information Sources . . . . . . . . . . . . . . . . . 15
   The QuEST Forum's Global Approach . . . . . . . . . . . . 16
   End Notes . . . . . . . . . . . . . . . . . . . . . . . . . 16

## Chapter 2  Common Requirements . . . . . . . . . . . . . . . . 19
Introduction . . . . . . . . . . . . . . . . . . . . . . . . . . . . . . 19
Codes Used to Identify the Type of Requirement . . . . . . . 20
Common Adders . . . . . . . . . . . . . . . . . . . . . . . . . . . 20
    4.1 Management Responsibility . . . . . . . . . . . . . 21
    4.2 Quality System . . . . . . . . . . . . . . . . . . . . . 21
    4.2.3 Quality Planning . . . . . . . . . . . . . . . . . . 22
    4.3 Contract Review . . . . . . . . . . . . . . . . . . . 25
    4.4 Design Control . . . . . . . . . . . . . . . . . . . . 26
    4.6 Purchasing . . . . . . . . . . . . . . . . . . . . . . . 35
    4.10 Inspection and Testing . . . . . . . . . . . . . . . 36
    4.13 Control of Nonconforming Product . . . . . . . . . 36
    4.14 Corrective and Preventive Action . . . . . . . . . . 37
    4.15 Handling, Storage, Packaging, Preservation,
        and Delivery . . . . . . . . . . . . . . . . . . . . . . 38
    4.18 Training . . . . . . . . . . . . . . . . . . . . . . . . 39
    4.19 Servicing . . . . . . . . . . . . . . . . . . . . . . . 42
    4.20 Statistical Techniques . . . . . . . . . . . . . . . . 44
    4.21 Quality Improvement and Customer Satisfaction . . 45
End Notes . . . . . . . . . . . . . . . . . . . . . . . . . . . . . . 49

## Chapter 3  Hardware, Software, and Service Requirements . . . . . . . . . . . . . . . . . . . . . . . . . . . . . 51
Introduction . . . . . . . . . . . . . . . . . . . . . . . . . . . . . . 51
Codes Used to Identify the Type of Requirement . . . . . . . 51
Hardware/Software Adders . . . . . . . . . . . . . . . . . . . . 52
    4.8 Product Delivery . . . . . . . . . . . . . . . . . . . 53
    4.19 Servicing . . . . . . . . . . . . . . . . . . . . . . . 54
Hardware/Services Adders . . . . . . . . . . . . . . . . . . . . 55
    4.4.8 Design Validation . . . . . . . . . . . . . . . . . . 55
    4.9 Process Control . . . . . . . . . . . . . . . . . . . . 55
    4.10 Inspection and Testing . . . . . . . . . . . . . . . 57
Hardware Adders . . . . . . . . . . . . . . . . . . . . . . . . . . 58
    4.4 Design Control . . . . . . . . . . . . . . . . . . . . 58
    4.8 Product Identification and Traceability . . . . . . . . 61
    4.9 Process Control . . . . . . . . . . . . . . . . . . . . 62
    4.10 Inspection and Testing . . . . . . . . . . . . . . . 62
    4.11 Control of Inspection, Measuring, and
        Test Equipment . . . . . . . . . . . . . . . . . . . . 63
    4.15 Handling, Storage, Packaging, Preservation,
        and Delivery . . . . . . . . . . . . . . . . . . . . . . 63

|     |     |
| --- | --- |
| 4.19 Servicing | 64 |
| 4.21 Quality Improvement and Customer Satisfaction | 64 |
| Software Adders | 65 |
| 4.2 Quality System | 65 |
| 4.4 Design Control | 66 |
| 4.5 Document and Data Control | 69 |
| 4.9 Process Control | 69 |
| 4.10 Inspection and Testing | 71 |
| 4.15 Handling, Storage, Packaging, Preservation, and Delivery | 71 |
| 4.19 Servicing | 72 |
| Service Adders | 73 |
| 4.4 Design Control | 74 |
| 4.9 Process Control | 75 |
| 4.21 Quality Improvement and Customer Satisfaction | 76 |
| End Notes | 76 |
| **Chapter 4  TL 9000 Metrics** | **77** |
| The Importance of Metrics | 77 |
| Introduction | 78 |
| Metrics Work Group | 79 |
| Metrics Selection Using the Goal-Question-Metric Model | 81 |
| Phase 1 Metrics | 84 |
| Metrics Usage | 86 |
| Common Metrics | 87 |
| Number of Problem Reports (NPR) | 87 |
| Fix Response Time (FRT) and Overdue Problem Report Fix Responsiveness (OFR) | 88 |
| On-Time Delivery (OTD) | 90 |
| System Outage | 92 |
| The Hardware Only Metric | 94 |
| Return Rate | 94 |
| Software Only Metrics | 96 |
| Software Update Quality (SWU) | 96 |
| Release Application Abort (RAA) | 98 |
| Corrective Patch Quality (CPQ) and Feature Patch Quality (FPQ) | 100 |
| The Service Only Metric | 102 |
| Service Quality | 102 |
| Metrics Submission and Analysis | 105 |
| End Notes | 105 |

# Table of Contents

**Chapter 5 TL 9000 Metrics Data Management and Analysis** ............................... **107**
    Metrics Repository System ..................... 107
        Confidentiality of Source and Data ........... 108
        Encryption Process ....................... 109
        Data Integrity ........................... 109
    Metrics Data Storage ......................... 110
    Metrics Data Flow ........................... 111
        Interaction Between Supplier and Customer ....... 114
        Interaction Between Supplier and ASQ ......... 115
        Interaction Between Supplier and UTD ......... 115
        Interaction Among Supplier, ASQ, and UTD
            when Data Transmission Fails .............. 116
        Interaction Among Suppliers, Registrar, and ASQ .... 117
        Interaction Between ASQ and UTD ........... 118
        ASQ's Additional Responsibilities ............ 118
    The Goals and Expectations of TL 9000 Metrics ....... 118
        The Use of Metrics to Improve Product and
            Service Quality ..................... 120
        Metrics Computation ..................... 121
        Product and Service Categories .............. 122
        Training ............................. 123
    The QuEST Forum's Global Approach ............ 124
    End Notes .................................. 124

**Chapter 6 TL 9000 Case Studies: An Innovative Way of Gathering Feedback Information** ................ **125**
    Introduction ............................... 125
    Pilot Participants ........................... 126
    Case Study Surveys ......................... 126
        The Case Study Survey Results .............. 127
        Some Facts About the Case Study
            Survey Respondents .................. 129
    Obtaining Management Commitment ............ 130
        Cost vs. Value ......................... 131
        Managing the Implementation Process .......... 135
        Benefits of the Gap Analysis and Internal Audits .... 136
    TL 9000 Process Development and Documentation ...... 137
        Training ............................. 138
        Internal Auditing ....................... 138
        Metrics .............................. 138
        The Lifecycle Model ..................... 139

Pre-Assessment and Registration Results . . . . . . . . . . . 140
Recommendations to Organizations Implementing
  TL 9000 . . . . . . . . . . . . . . . . . . . . . . . . . . . 141
End Notes . . . . . . . . . . . . . . . . . . . . . . . . . . . 142

**Chapter 7  Some Final Thoughts** . . . . . . . . . . . . . . . . **143**

**Appendix 1  Table of Adders to ANSI/ISO/ASQC Q9001–
1994 and ANSI/ISO/ASQ Q9001–2000** . . . . . . . . . . . . . . **145**

**Appendix 2  TL 9000 Registration and Accreditation** . . . . . **153**
Introduction . . . . . . . . . . . . . . . . . . . . . . . . . . 153
Getting Started . . . . . . . . . . . . . . . . . . . . . . . . 153
  The Initial Audit . . . . . . . . . . . . . . . . . . . . . 154
  Migration to TL 9000 . . . . . . . . . . . . . . . . . . . 155
  Publicizing Your Registration . . . . . . . . . . . . . . 155
  Your Registrar . . . . . . . . . . . . . . . . . . . . . . 155
  Alternate Method for Maintaining Certification . . . . . 156
Conclusion . . . . . . . . . . . . . . . . . . . . . . . . . . . 157

**Bibliography** . . . . . . . . . . . . . . . . . . . . . . . . . **159**

**Index** . . . . . . . . . . . . . . . . . . . . . . . . . . . . . **161**

# Foreword

The telecommunications industry is globalizing, restructuring, and growing dramatically to become the backbone of the new digital economy. Some of the world's largest companies are in a headlong competition to provide a high-speed digital world connection to customers, either by telephone lines, cable TV connections, wireless or satellite connections. The quality and reliability of these new digital networks, and the supply lines required to build them, must continually improve to deliver on the full promise of the new technology on a global basis. The QuEST Forum, a collaborative effort between the telecommunications service providers and their suppliers, addressed this challenge by developing and deploying the TL 9000 Quality System focused on improving the performance of the new digital telecom industry.

TL 9000 is about improving performance: better overall product quality, lower costs, reduced cycle time, and improved customer satisfaction. The requirements and metrics build on currently used industry standards, including the ISO 9000 series, and is rapidly gaining worldwide momentum and acceptance. It's specific to Telecom, which creates the opportunity to measure and drive industry improvement.

This book captures and interprets the TL 9000 Quality System requirements, and cost and performance-based metrics. It provides a step-by-step approach by first discussing the standard, describing the rationale for the existence of the standard, and then sharing tips on how to meet the requirements. The authors have been participants in the QuEST Forum from its inception and have worked diligently to write this book. They show us that the solution for better quality is not just testing, inspections, or documented procedures but requires a broad and realistic view of customer requirements and how products and services can affect these requirements. We are reminded that successful processes are vital to the reputation and health of an organization.

As you read this book, be prepared to be challenged by new concepts that can produce improved software, hardware, service quality, and satisfied and loyal customers. It will be an important tool to aid in the improvement of telecommunications products and services in any company choosing to use it.

Stephen G. Welch
President
SBC Procurement
QuEST Forum Founding Chairman

# Preface

The Quality Excellence for Suppliers of Telecommunications (QuEST) Leadership Forum was formed as a telecommunications sector consortium in January 1998. The QuEST Forum launched a worldwide supply-chain initiative to develop quality management system standards aimed at improving the products and services provided by the industry. The dynamic nature of telecommunications and the globalization of the industry highlighted a need for a single set of telecommunications quality management system standards. The Forum brought together service providers and suppliers early in 1998 and produced the first two handbooks, *TL 9000 Quality Systems Requirements* and *TL 9000 Quality Systems Metrics*, by the end of 1999.

The industry makes annual purchases estimated at more than $125 billion worldwide. Of this, the cost of poor quality is estimated to be $10–15 billion. The potential savings from implementing TL 9000 is therefore considerable. This is especially important since it is estimated that the standards will be applicable to over 10,000 suppliers worldwide.

One major new aspect of TL 9000 is the metrics for hardware, software, and service quality. It is expected that the metrics will be a major benchmarking tool that will help the industry improve quality and, hence, drive down cost. Historically the industry has benefited from hardware quality improvement. Between 1985 and 1996, the annual cost of poor quality (COPQ) of hardware within the U.S. telecommunications industry was measured at $2.5 billion; but as the products improved, the annual COPQ was reduced to $750 million.

The three authors of this book have worked on TL 9000 from the beginning and bring to the table strong backgrounds in the hardware, software, and service aspects of telecommunications. They are also well versed in the issues debated during the development of TL 9000, the industry need to develop additional requirements beyond ISO 9001 [1], and the identification of metrics that are aimed at driving continual improvement. They also have experience implementing TL 9000 in their organizations.

The Forum started with 44 member companies of the telecommunications industry. As of December 31, 2000 there were 151 members, including 14 service providers, 100 suppliers, and 37 liaison members. The goal to have a membership of 150 by the end of 2000 has been reached.

This book presents information in seven chapters. The introductory chapter covers the creation of QuEST Forum, its membership categories, the two handbooks, key differences between ANSI/ISO/ASQC Q9001–1994 and TL 9000, the registration process, and the global growth initiative. Chapter 2 describes the additional common requirements (adders) of TL 9000. The adders are highlighted, explained, and "tips" on implementation are provided. Chapter 3 covers the same material for the hardware, software, and service adders. Chapter 4 covers the individual metrics and describes their purposes. Chapter 5 describes the metrics collection process, analysis of the data, and the product categories. Chapter 6 discusses the case studies and data received from the pilot projects. Finally, chapter 7 concludes with the benefits of TL 9000 implementation and future activities of the QuEST Forum. The appendices provide a table of adders in terms of ANSI/ISO/ASQC Q9001–1994 and ANSI/ISO/ASQ Q9001–2000 and a description of the accreditation/registration process. Each chapter contains a list of references. A bibliography is provided at the end of the book.

# Acknowledgments

The authors would like to thank all the past and present members of the QuEST Forum whose support made it possible for us to write this book. We would especially like to thank Steve Welch, President, SBC Corporate and Administrative Services for his support and the Foreword to this book; Jim Mroz of *The Informed Outlook*, for contributing to the development of chapter 6 on TL 9000 Case Studies; and Dawn Perry of Cisco Systems for contributing the appendix on "Accreditation and Registration."

We would like to thank our editors, Annemieke Koudstaal and Ken Zielske, for their support and counsel during the development of the text. In addition, we would like to acknowledge the support of the following individuals:

| Name | Company | Support |
| --- | --- | --- |
| Reg Blake | BSI | Pilot Survey |
| Bob Brigham | Telcordia Technologies | Seed documents and planning the structure of the book |
| Kevin Calhoun & Sandy Holston | Siecor | Pilot Survey |
| Don Carlson & Art Morrical | Lucent Technologies | Pilot Survey |
| Michael Faccone | Telcordia | Pilot Survey |
| Rolf Hendrickson | Nortel Networks | Pilot Survey |
| Wally Juban | Pirelli Cables | Pilot Survey |
| Don Moore | Fujitsu | Pilot Survey |
| Malcolm Phipps | QMI | Pilot Survey |
| Gary Reams | NEC America | Pilot Survey |
| Donna Reinsch | Marconi | Pilot Survey |
| Pierre Salle | KEMA | Pilot Survey |
| Joel Savilonis | NSAI | Pilot Survey |

| Name | Company | Support |
|---|---|---|
| Joe Taylor | Tellabs | Pilot Survey |
| Jim Woodard | SBC—CA | Pilot Survey |
| Rose Hoff | Advanced Fibre Communications (AFC) | Reviewer |
| Zvi Ravia | Comverse | Reviewer |
| John Walz | Lucent Technologies | Reviewer |
| Rick Werth | SBC | Reviewer |
| Tom Yohe | Alcatel | Reviewer |
| Richard Morrow | UTD | Reviewer |

# 1

# Introduction

## CREATION OF THE QUEST FORUM

In early 1996, Bell Atlantic, BellSouth, Pacific Bell (later acquired by SBC), and Southwestern Bell (SBC), executives initiated an effort to establish better quality management requirements for the industry. They represented the future expectations of the leading telecommunications service providers in the United States.

After a year and a half of planning, the four service providers invited their leading suppliers to a meeting in Baltimore in October 1997. The result was the formation of a cooperative venture among the service providers and suppliers to create an industry quality management system standard aimed at improving the quality of telecommunications products and services.

Work began in January 1998 to draft a set of quality requirements to be incorporated into a TL 9000 Handbook. To accomplish this, work groups consisting of industry experts in software, hardware and services were formed. The work groups were chartered with the development of the TL 9000 hardware, software, and service requirements to be added to the ISO 9001[1] requirements. These added requirements, or "adders," would be based on the historical needs of the industry. Additional work groups were formed to develop a second handbook consisting of hardware, software, and service metrics. A regular monthly schedule of team meetings was initiated with the goal of having a working implementation program by the end of 1999.

It was understood from the beginning that cost and performance-based metrics were especially important to the effort. This is because they can be used to quantify the benefits gained, assess progress in quality maturity, and identify areas where the quality process improvement will have the greatest cost impact and provide comparative benchmarking capabilities for the industry.

The QuEST Forum recognized the parallel between what they wanted to accomplish and what had been done by the automotive industry a few years earlier. The auto industry had created a standard based on ISO 9001 with adders called QS-9000.[2] The Forum consulted with the auto industry forum, The Automotive Industry Action Group (AIAG), during the start-up phase.

## TELECOMMUNICATIONS INDUSTRY NEEDS

The working groups reviewed ISO 9001 and determined that supplemental requirements were needed in the following areas:

- *Reliability and associated costs.* The industry places a great emphasis on reliability because of end-user demands placed on service providers. The networks, switching systems, and local transmission equipment must operate continuously to accurately pass information between end users. In this high technology information age, the rapid transfer of data requires highly reliable products and services.

- *Software development and life cycle management.* Software is an integral part of telecommunications product and service offerings. Although ANSI/ISO/ASQ Q9000-3–1997[3] is a guide for applying ANSI/ISO/ASQC Q9001–1994 to software quality management systems, it does not cover the software lifecycle needs of the industry. TL 9000 developers used inputs from ANSI/ISO/ASQ Q9000-3–1997,[3] ISO/IEC 12207,[4] TR-179[5] and other sources to satisfy most of these needs.

- *Specialized service functions.* Installation, engineering, maintenance, repair, call center, and support services are integral to the customer-supplier relationship in the telecommunications industry. In addition the coverage of servicing in clause 4.19 in ANSI/ISO/ASQC Q9001–1994[1] is extremely limited and additional requirements are needed to provide "after market" functions such as emergency service, problem notification, and supplier support.

- *Continuation and further development of the relationships between service providers and suppliers.* A unique characteristic of the telecommunications industry is the ongoing relationship between service providers and their suppliers. Service providers require new or improved features on a regular basis. Thus there is a need for continual communication between customers and their suppliers.

These industry needs led to the development of the 83 adders and the 11 metrics defined in the two TL 9000 Handbooks.

# THE GOALS AND EXPECTATIONS OF TL 9000

There were five major goals defined by the QuEST Forum:

1. To foster systems that will protect the integrity of telecommunications products, services and networks.
2. To develop requirements that will help organizations more accurately assess the implementation of their quality management system.
3. To identify tools that will drive continual improvement of products, services, and processes.
4. To develop standard metrics for use as a continual improvement tool.
5. To implement an industry standard assessment process that will reduce the multiplicity of conflicting programs.

There are a number of industry benefits that the QuEST Forum membership expects from the development of TL 9000. These include:[6]

- Improved service to end users;
- enhanced customer-supplier relationships (Note: The QuEST Forum was designed as a cooperative effort among customers and suppliers);
- a reduction in the number of external 2nd and 3rd party audits and site visits;
- uniform performance and cost-based metrics for use as benchmarks for improving product and service quality;
- enhanced supply chain management, including 2nd and 3rd tier suppliers; and
- the creation of a platform for industry improvement initiatives.

Specific expectations for Forum member products and services include:

- cycle-time reduction
- improved on-time delivery
- reduced life-cycle costs
- superior products
- defect reduction
- increased profitability and market share

## QuEST Forum Membership Categories

There are five classes of membership in the QuEST Forum:

1. Service providers
2. Suppliers
3. Liaison members
4. Non-member liaisons
5. Recognized consultants

Service providers and suppliers are full members with full privileges and responsibilities. The privileges include voting rights, access to data, and full participation in working groups. The responsibilities include full payment of dues (currently $10,000 per year), a requirement to provide at least one subject matter expert for the working groups, and attendance at a specified number of meetings.

Liaison members come from a number of communities such as the accreditation agencies, registrars, training organizations, consultants, other standards bodies, and other industry organizations. They do not have voting privileges or access to the data and may be restricted from membership in certain working groups. Their responsibilities include a reduced payment of dues (currently $2000), and a prohibition against commercial activities while attending official Forum meetings. Each liaison member must be sponsored by at least two full members.

Non-member liaisons are organizations that have a specific expertise or represent a body of knowledge and/or members that provide a resource critical to the Forum's future growth and expansion of the TL 9000 programs. An example of a non-member liaison is the International Accreditation Forum. There are no membership fees and they are limited to specific activities that utilize their expertise.

Recognized consultants are associated with the Forum, but cannot be members. To be recognized, they must attend a QuEST Forum sanctioned TL 9000 Requirements and Metrics course, pass the written exam, and attend at least one "Consultant Information Exchange" session per year. They must also agree to have their TL 9000 clients complete a Forum administered Customer Satisfaction Survey, adhere to the QuEST Forum Representatives Guiding Principles, and agree to resolve any conflicts or problems by using the QuEST Forum mediation board. Consultants must be sponsored by at least two full members.

The major benefits for both the full and liaison members are:

- *Participation in the development of the quality requirements.* A company that joins the Forum as a full or liaison member has access to and helps develop TL 9000 Requirements and Metrics Handbooks. The results for member companies should be:
    - Increased efficiency and productivity as a result of their efforts in developing the TL 9000 handbooks.
    - More uniform internal and external audits.
- *Voting rights.* Full member: one vote per member company on all decisions. Liaison members cannot vote.
    - Thus, member companies play a part in developing consensus quality system requirements, metrics, and by-laws.
- *Industry data for all product categories.* Viewing privileges for all of the aggregate industry data.
    - Full members get access to valuable benchmarking industry data for use in comparing and improving products and services.
    - Liaison members do not have access to the data, but they may purchase the data.
- *Work group interaction.* Full members must provide at least one subject matter expert to participate in work group activities. However, they may provide as many subject matter experts to work on as many work groups as they desire.
    - Members receive valuable information and understanding of requirements and metrics and have a voice in shaping them.
- *Forum meetings held three or four times per year.* Members can send representatives to Forum meetings.
    - Members hear informative speakers, receive updates from the Executive Board, participate in workshops, and establish key networking contacts.
- *Forum member discounts.*
    - Members receive discounts on Forum products such as the handbooks, Forum merchandise, and training materials. Site licenses have been developed for the training courses and for obtaining copies of the handbooks at reduced prices.

## The TL 9000 Model

The basic foundation of the TL 9000 model is ANSI/ISO/ASQC Q9001–1994. TL 9000 incorporates the 20 elements of that standard and adds 83 new requirements, or "adders," including the additional element 4.21,

**Figure 1.1** The TL 9000 Model.[6]
(Reprinted with permission of the QuEST Forum.)

"Quality Improvement and Customer Satisfaction." The adders are categorized as common requirements, hardware specific requirements, software specific requirements, service specific requirements or paired combinations of specific requirements.

A second set of requirements covers the metrics. Again, these are categorized into common, hardware specific, software specific, and services specific metrics.

Figure 1.1 depicts the TL 9000 model.

### The TL 9000 Handbooks

The Forum developed two handbooks during 1998 and 1999. Book One[6] contains the verbatim ISO 9001 requirements, the 83 adders, and appendices covering accreditation, registration, and guidance for customer-supplier communication. Book Two[7] contains a description of the metrics, responsibilities of all parties with respect to the metrics, and the product category and normalization tables. Updated Product Category Tables can be found on the QuEST Forum web site (www.questforum.org).

## SOURCES USED TO DEVELOP THE TL 9000 REQUIREMENTS HANDBOOK

At the start of the development of TL 9000 there were many industry documents to use as sources. The following are the publications that were chosen as sources for the adders:

- ANSI/ISO/ASQC Q9001–1994,[1] the basic quality management standard. Section 4 of this document contains requirements which were included verbatim in TL 9000, Book One;

- ISO 9000-3,[3] the ISO document that provides guidance for applying ISO 9001 to software;

- ISO/IEC 12207,[4] the international standard that describes quality management system software life cycle requirements;

- TR-NTW-000179,[5] the Telcordia Technologies document that describes telecommunications software requirements;

- GR-1202,[8] the Telcordia Technologies document that describes the Telcordia "Customer Supplier Quality Process;"

- GR-1252,[9] the Telcordia Technologies document that describes telecommunications hardware requirements;

- ISO 9004-2,[10] the ISO document that provides quality management system guidelines for service organizations.

In addition, the QuEST Forum membership provided an added set of requirements and QS 9000,[2] ISO 10011-1,[11] ISO 10011-2,[12] and ISO 10011-3[13] provided inputs for the appendices.

Table 1.1 contains a summary of the number of contributions to the Requirements Handbook[6] from each source. GR-1252, the Telcordia Technologies hardware requirements document, provided the largest number of inputs. However, three software requirements documents, TR-179, ISO/IEC 12207 and ISO 9000-3, collectively provided 32 requirements or parts of requirements.

**Table 1.1** The Sources of the Adders.

| Source | Number of Requirements | Number of Notes |
|---|---|---|
| GR-1202 | 16 | 0 |
| GR-1252 | 23* | 0 |
| TR-179 | 13* | 0 |
| ISO 12207 | 13** | 1 |
| ISO 9000-3 | 6 | 1 |
| ISO 9004-2 | 1 | 0 |
| QuEST Forum Membership | 13 | 0 |
| Total | 83^ | 2 |

Notes:
\* This source contributed to one requirement jointly with one other source.
\*\* This source contributed to two requirements jointly with one other source.
^ Two requirements were developed from multiple sources (two sources in each case).

# KEY ADDITIONS TO ANSI/ISO/ASQC Q9001–1994

The following areas are where the QuEST Forum focused its attention and expanded on the current ISO 9001 requirements:

- More emphasis on top management responsibilities.
- More focus on proper/robust planning:
  - quality planning
  - project planning
  - configuration management planning
  - product planning
  - life cycle planning
  - test planning
- Requirements for customer-supplier communication.
- Requirements emphasizing quality improvement and customer satisfaction.
- Requirements covering specialized service functions.
- Metrics required for certification:
  - objectives include a requirement to define metrics targets
  - metrics must be defined, tracked and reported to a central data base
  - metrics must be used to foster continual quality improvement
  - auditors responsibilities with respect to metrics:
    - assuring the suppliers' metrics processes are working
    - assuring the results are used by the suppliers to improve products, services, and processes
- Unlike ISO 9001: *should* means the preferred approach
  - suppliers choosing other approaches must be able to demonstrate that their approach meets the intent of TL 9000
- Training sanctioned by the QuEST Forum and delivered by two authorized training providers. Currently the course providers are the Excel Partnership, and STAT-A-MATRIX (The SAM Group).

# THE USE OF METRICS TO IMPROVE PRODUCT AND SERVICE QUALITY

The unique difference between TL 9000 and the current ISO 9001 practice is the use of metrics in continual quality improvement. These are measurements used to determine the quality level of product and services. Examples are circuit pack return rates, number of problem reports, and software update quality. It is expected that extending ISO 9000 in this direction will result in an industry drive to improve overall quality. Certainly, when suppliers know where their products stand relative to their competitors, and their customers know the quality level of "best in industry," there will be a drive by all to improve.

The QuEST Forum envisions multiple uses for the metrics. For example, they can be used to enhance customer-supplier communication, prioritize and solve the most costly problems, and bridge the gap between quality issues and business results. Some service providers have indicated that they will use the metrics to develop report cards for their suppliers.

The Forum has retained the University of Texas at Dallas (UTD) to gather the data and develop statistics for each product class. The statistics will identify the mean (or median) standard deviation (or range) and "best in industry." The data will be published on the QuEST Forum web site (www.questforum.org). UTD was selected because of its long-standing role in telecommunications engineering education. It's expected that UTD will provide insight into benchmarking industry quality problems, identifying quality trends, and identifying possible solutions to quality problems.

## The Metrics Implementation Process

The supplier gathers the data on a monthly basis and sends three months of data to UTD for inclusion in the industry database. A double blind system has been established to insure that the data is secure and not available to competitors or other unauthorized persons. The process works as follows:

- When a supplier has gathered the data, a notification is sent to the Forum Administrator, The American Society for Quality (ASQ), who provides encryption keys back to the supplier to be used in sending the data to the industry database.

- The supplier encrypts the data and sends it to the Metrics Repository System (MRS) at UTD. Personnel transfer the data from the receiving server into the industry database. At least two individuals must be present during the transfer to ensure security of the data.

A major concern of the QuEST Forum membership has been the security of the data being supplied to UTD. The double blind process ensures that the data will be safe from access by competitors. In addition, no single individual at UTD will be able to access the data alone. The whole process has been developed with a minimum cost to suppliers.

UTD has complied with BS 7799,[14] the information security management requirement document developed by the British Standards Institute (BSI). UTD is the first U.S. organization to be certified to this document by BSI.

## Approved Metrics

Eleven metrics in four categories have been approved. These are as follows:[7]

1. Common to hardware, software, and services:
   - Number of Problem Reports
   - Problem Report Fix Response Time
   - Overdue Fix Responsiveness
   - System Outage Measurement
   - On-time delivery

2. Hardware only:
   - Return Rates

3. Software only:
   - Corrective Patch Quality
   - Feature Patch Quality
   - Software Update Quality
   - Release Application Aborts

4. Service Only:
   - Service Quality

A major issue during the development of these metrics was the identification of *normalization factors* which allow comparison of similar products with different characteristics. For example, for circuit switching, the normalization factor for return rate is returns/10,000 terminations/year. Thus products with a large number of terminations per circuit pack are measured fairly against those with fewer terminations per circuit pack.

Another issue was a result of a requirement of some service providers that the suppliers provide Reliability and Quality Measurements for Telecommunications Systems (RQMS) metrics defined in the Telcordia Technologies document GR-929-CORE.[15] In some cases the supplier may provide the RQMS metrics in place of the corresponding TL 9000 metrics.

Other metrics and indicators are under consideration, but have not been approved. Indicators are a separate category of measures that are used to "flag potential cost, schedule, productivity, and quality issues," but are not reportable to UTD.[16]

The registrar has an important role in assuring compliance to the metrics requirements. "This includes assuring data validity and integrity, verifying [that] the benchmark metrics have been reported within the required time frame, and [assuring] that any metric-specific nonconformance is resolved in a timely manner."[17] Also, because of the requirement in Book One that "objectives for quality shall include targets for the TL 9000 metrics defined in the *TL 9000 Quality System Metrics Handbook*,"[6] the registrar has a role in assuring this use of the metrics. Finally, there are requirements in reference six for the supplier to collect and analyze customer satisfaction, field performance, and service performance data, some of which is metrics data.[6]

## PRODUCT AND SERVICE CATEGORIES

The initial set of product and service categories was provided in Appendix A of the Metrics Handbook.[7] Official Product and Service categories are subject to change. Because of the possibility of change, they are stored on the QuEST Forum web site (www.questforum.org). The following are the product and service categories defined for the metrics implementation as of June 2000:[7]

- *Switching equipment* broadly defined to include packet and circuit switched architectures
- *Signaling equipment* consists of five basic categories: supervisory, information, address, control, and alerting
- *Transmission equipment* connects the switched and interoffice networks with individual customers
- *Operations and Maintenance equipment* used to manage communications networks
- *Common Systems* including power systems and Network Equipment-Building System (NEBS) that provide space and environmental support for the network[18]
- *Customer premises equipment* commonly installed at the subscriber's location
- *Services* which include installation, engineering, maintenance, repair, call center, and support
- *Components and Sub-assemblies* including component suppliers, contract manufacturers, and OEM suppliers

## PILOT PROGRAM

One of the QuEST Forum initiatives was the creation of a pilot program to validate TL 9000 and its associated activities. The goals of the program were to provide feedback on:

- the bottom-line value of the "adders;"
- the processes that form the metrics reporting system;
- the processes used to manage the registrar accreditation and supplier assessment/certification system;
- the sanctioned training courses; and
- the auditability of the requirements and measurements.

Feedback was accomplished through weekly conference calls which included minutes posted on the web site (www.questforum.org), monthly meetings, documented lessons learned, case studies, and commonly asked questions posted on the web site.

Sixteen organizations from eleven companies participated in the program in four categories. Eight were hardware suppliers, three were software suppliers, four combined hardware and software, and one was a supplier of services. The participants are shown in Table 6.1, chapter 6.

The first step for each pilot organization was to make arrangements for its registrar to participate in the pilot program. Each registrar had to have its auditors take the QuEST Forum sanctioned training and pass the auditor exam. Also, at least one member of the registrar's certification board with veto power or a majority of the board had to take the training and pass the exam. The registrar's accreditation agency also had to have at least one member of the registration decision-making body take the QuEST Forum sanctioned training and pass the written exam. In addition, the accreditation agency had to witness at least one audit performed by each registrar using auditors who have completed the training and passed the exam.

Each pilot organization had to train its auditors and implementers, perform a gap analysis, develop processes for collection of the metrics, and submit the metrics data to UTD. The organizations that completed the process and were certified were not allowed to advertise the fact until the end of January 2000. They were honored at the January 25, 2000 Forum meeting in Dallas, Texas.

## REGISTRATION PROCESS

The formal registration process started in the first quarter of 2000. The QuEST Forum established requirements for accreditation bodies to qualify registrars who will carry out TL 9000 certifications.[6] A supplier must demonstrate conformance to TL 9000 by successfully completing a third-party certification audit from an accredited TL 9000 registrar. The supplier will obtain a certificate in any combination of hardware, software, and/or services:

- Hardware: TL 9000-HW
- Software: TL 9000-SW
- Service: TL 9000-SC

Note that the supplier must either be ISO 9001 or 9002 certified or must include certification to either of these standards as part of the TL 9000 certification.

The QuEST Forum has been working with the Registrar Accreditation Board (RAB), the organization that oversees the competency of Quality Management Systems and Environmental Management Systems certification bodies in the United States, and the Standards Council of Canada (SCC) to pilot the procedures. Other accreditation agencies, such as the members of the International Accreditation Forum, will be invited to participate in the future. The following are the web sites for these organizations:

- RAB (www.rabnet.com)
- SCC (www.scc.ca)

## TRAINING

The QuEST Forum has authorized two organizations, STAT-A-MATRIX (The SAM Group) and Excel Partnership to develop and provide sanctioned TL 9000 training. The Forum retains the rights to the training intellectual property. The materials have been translated into French, Spanish, German, and Japanese. Translation into Chinese is in progress.

By selecting two training organizations to jointly provide training, the Forum expanded its philosophy of having competitors cooperate for the good of the industry. The two organizations should be commended for their joint effort and sharing of responsibilities and products. Also, the Forum should be commended for its foresight in taking this approach to providing sanctioned training. The results were better because two different perspectives were blended into the courses, making them stronger than the courses each trainer could have provided separately.

The two training providers have developed four sanctioned courses:

1. TL 9000 Quality System Overview: A half-day course aimed at the majority of company associates, including senior management, which describes the goals of TL 9000 and how it can aid the organization.

2. TL 9000 Quality System Implementation: A three day course that provides a road map for successful TL 9000 implementation and compliance.

3. TL 9000 Quality System Requirements and Metrics Auditing: A four day course that provides auditor training on the quality management system "adders" and the metrics requirements. A written exam is given at the completion of the course. Passage of the course and exam is required to be certified as a TL 9000 auditor.

4. TL 9000 Quality System Auditing for Registrars: A three day course designed for auditors from registrar organizations. A test is given at the completion of the course. Passage is required to be certified as a TL 9000 registrar auditor.

Excel Partnership is providing a fifth sanctioned course, TL 9000 Metrics Data Submission. This course describes the details of submitting data to the database at UTD. The pilot organizations took early versions of the courses as part of the overall pilot program. They provided valuable input to the course developers that were later used to improve the courses.

## ADMINISTRATION

An Executive Board consisting of six service providers and six suppliers governs the Forum. The Chairmanship rotates on an annual basis between service providers and suppliers. The first chairman was Steve Welch, President, SBC Corporate and Administrative Services. Krish Prabhu, President and Chief Operating Officer, Alcatel Telecom USA, succeeded him in 2000. The Forum selected the American Society for Quality (ASQ) to be the Forum Administrator and the University of Texas at Dallas to administer the Metrics Repository System (MRS).

One reason for the selection of ASQ was its historical role in support of quality standards development that includes providing the secretariat for the United States Technical Advisory Groups to ISO/TC 176, ISO/TC 207, ISO/TC 69 and IEC/TC 56. Also, ASQ's role as a worldwide leader in the quality profession played an important part in the selection.

ASQ is responsible for many functions, including the following:[19]

- Performance of general business functions.

- Management of the program, including: QuEST Forum and working meetings; QuEST Forum board meetings and conference calls; membership and web site administration.

- Management of accreditation bodies and registrar information.

- Administration of membership responsibilities.

- Distribution and publishing of marketing materials.

- Administration of training and education responsibilities.

- Publication and distribution of the Quality System Requirements and Metrics Handbooks and promotion of their use.

- Arrangement of technical conferences and workshop programs.

The Erik Jonsson School of Engineering and Computer Science at the University of Texas at Dallas was selected to administer the Metrics Repository System. The responsibilities include storage, security, maintenance, and statistical analysis of the data. It is also expected that the university will provide research into the solutions of industry problems identified as a result of the analysis.

## Information Sources

The Forum's website (www.questforum.org) is the main source of information. The public section of the web site contains information on membership, meetings, and ordering books, materials, and merchandise. The member section contains the metrics database summary results, individual membership information, presentations, minutes, product and service categories, and other QuEST Forum proprietary materials.

## THE QuEST FORUM'S GLOBAL APPROACH

The Forum has made great strides in its goal to become a truly global telecommunications organization. It is the only global organization dealing with telecommunications quality management. Membership has grown internationally and includes service providers from the United Kingdom, South Africa, Australia, and Canada; suppliers from the United Kingdom, France, Israel, Canada, and many multinational suppliers based in the United States. In addition, the QuEST Forum has established liaison status with organizations in Europe, Russia, Japan, Korea, and Singapore.

The Forum has Liaison A status with ISO/TC 176, IEC/TC 56 and ISO/IEC JTC1/SC7. As a liaison member of ISO/TC 176, the Forum has provided four members of task groups working on the year 2000 revisions to ANSI/ISO/ASQC Q9001–1994 and ANSI/ISO/ASQC Q9004–1994. These individuals have contributed to the verification, validation, transition, and implementation processes used by ISO/TC 176.

The Forum has also developed a cooperative relationship with the EIRUS (European In-Process Quality Metrics [IPQM] and RQMS User Group) organization, a European telecommunication metrics user group. EIRUS uses two sets of metrics based on Telecordia Technologies requirements: European In-Process Quality Metrics (E-IPQM), and European Reliability and Quality Measurements for Telecommunications Systems (E-RQMS).[20] Joint working groups have been established to align the EIRUS and QuEST Forum metrics.

The QuEST Forum held Forum meetings in Brussels in June 1999, Paris in May 2000, Tokyo in June 2000, and London in October 2000. Future meetings will be scheduled for other parts of the world.

## END NOTES

1. ANSI/ISO/ASQC Q9001–1994, *Quality Systems—Model for Quality Assurance in Design/Development, Production, Installation and Servicing*, Second Edition (Milwaukee, WI: American Society for Quality Control, 1994).
2. *Quality System Requirements, QS-9000,* Third Edition (Detroit, Michigan: Automotive Industry Action Group [AIAG], March 1998).
3. ANSI/ISO/ASQ Q9000-3–1997, *Quality Management and Quality Assurance Standards—Part 3: Guidelines for the Application of ISO 9001:1994 to the Development, Supply, Installation, and Maintenance of Computer Software*, Second Edition (Milwaukee, WI: American Society for Quality, 1997).

4. ISO/IEC 12207, *Information Technology Software Life Cycle Processes* (Geneva, Switzerland: International Organization for Standardization, February 1995).
5. Telcordia Technologies, TR-NWT-179, *Quality System Generic Requirements for Software* (Morristown, NJ: Telcordia Technologies, Issue 2, June 1993).
6. *TL 9000 Quality System Requirements*, Book One, Release 2.5 (The QuEST Forum, 1999).
7. *TL 9000 Quality System Metrics*, Book Two, Release 2.5 (The QuEST Forum, 1999).
8. GR-1202-CORE, *Generic Requirements for Customer Sensitive Quality Infrastructure* (Morristown, NJ: Telcordia Technologies, Issue 1, October 1995).
9. GR-1252, *Quality System Generic Requirements for Hardware* (Morristown, NJ: Telcordia Technologies, Issue 1, May 1995).
10. ANSI/ISO/ASQC Q9004-2–1994, *Quality Management and Quality System Elements—Part 2: Guidelines for Services* (Milwaukee, WI: American Society for Quality Control, 1994).
11. ANSI/ISO/ASQC Q10011-1–1994, *Guidelines for Auditing Quality Systems—Auditing* (Milwaukee, WI: American Society for Quality Control, 1994).
12. ANSI/ISO/ASQC Q10011-2–1994, *Guidelines for Auditing Quality Systems—Qualification Criteria for Quality Systems Auditors* (Milwaukee, WI: American Society for Quality Control, 1994).
13. ANSI/ISO/ASQC Q10011-3–1994, *Guidelines for Auditing Quality Systems—Management of Audit Programs* (Milwaukee, WI: American Society for Quality Control, 1994).
14. BS 7799: 1999, *Information Security Management, Part 1: Code of Practice for Information Security Management and Part 2: Specification for Information Security Management* (London: The British Standards Institute, 1999).
15. GR-929-CORE, *Reliability and Quality Measurements for Telecommunications Systems (RQMS)* (Morristown, NJ: Telcordia Technologies, Issue 4, December 1998).
16. Henry Malec. "TL 9000 Database Repository and Metrics." *The Informed Outlook*, Vol. 4, No. 6 (June 1999): 4.
17. Galen Aycock, Jean-Normand Drouin and Thomas Yohe. "TL 9000 Performance Metrics to Drive Improvement." *ASQ Quality Progress* 32, no. 7 (July 1999): 41–45.
18. GR-63-CORE, Network *Equipment-Building System (NEBS) Requirements: Physical Protection* (Morristown, NJ: Telcordia Technologies, Issue 1, October 1995).
19. "Development of ASQ Role as TL 9000 Forum Administrator." *The Informed Outlook*, Vol. 4, no. 5 (May 1999): 1–2.
20. "Reliability Engineering." *European Quality Metrics: E-IPQM and E-RQMS*, Deliverable 5, EURESCOM P307 (November 1995). EURSECOM GmbH, Scloss Wolfsbrunnenweg 35, 69118 Heidelberg, Germany, www.eurescom.de/.

# 2
# Common Requirements

## INTRODUCTION

The TL 9000 requirements are based on ANSI/ISO/ASQC Q9001–1994[1] and 83 additional requirements or "adders," which are included to either enhance the quality system or be specific to the telecommunications industry. With the upcoming ANSI/ISO/ASQ Q9001–2000, consideration was given to these revised standards while developing the requirements of TL 9000. Some of the requirements of ANSI/ISO/ASQ Q9001–2000 are addressed in TL 9000.

The requirements are described in this chapter and chapter 3. In this chapter we discuss the requirements that are common to hardware, software, and service organizations. Chapter 3 contains a discussion of hardware only, software only, service only, and paired combinations of hardware, software, and service requirements. In both chapters we repeat the requirements from the requirements handbook[*],[2] give a rationale for the requirement, and provide "tips" from organizations that have implemented TL 9000. The reader should focus on the tips because they provide real information on how some organizations implemented the requirements.

The standard definition for the word *shall* in ANSI/ISO/ASQC Q9001–1994 is used in TL 9000, but *should* has a different meaning. The word *shall* is used for mandatory requirements that must be adhered to; *should* on the other hand indicates preferred practices. Suppliers choosing other approaches must be able to demonstrate that their approach meets the intent of TL 9000.

---

[*]The adders and notes in chapters 2 and 3 are repeated verbatim with permission of the American Society for Quality, publisher of *TL 9000 Quality System Requirements* for the QuEST Forum.

## CODES USED TO IDENTIFY THE TYPE OF REQUIREMENT

Individual codes were given to identify each adder as hardware, software, service specific, or when the adder is common to hardware, service, or software.

C: Common (hardware, software, and service)

HS: Hardware and software

HV: Hardware and service

H: Hardware only

S: Software only

V: Service only

## COMMON ADDERS

Common adders affect 10 elements. The number of adders in each element is shown in parenthesis. There are a total of 37 common adders:

4.1: Management responsibility (1)

4.2: Quality system (5)

4.4: Design control (8)

4.6: Purchasing (1)

4.13: Control of nonconforming product (1)

4.15: Handling, storage, packaging, preservation, and delivery (2)

4.18: Training (6)

4.19: Servicing (5)

4.20: Statistical Techniques (1)

4.21: Quality Improvement and customer satisfaction (7)

## 4.1 Management Responsibility

> **4.1.1.C.1**
> *Quality objectives*

 **Adder: 4.1.1.C.1: Quality Objectives:** Objectives for quality shall include targets for the TL 9000 metrics defined in the TL 9000 Quality System Metrics Handbook.

 **Rationale:** "If you don't know where you are going, how do you know you have arrived?" (Jarvis, Crandall, 1997[3]). In order to implement any process improvement you need to know where you are today. For example, if you start a diet to lose 20 pounds, you need to know your current weight—the baseline—before you begin the diet. The purpose of this baseline number is to learn whether or not the diet is working for you. The same principle applies to the quality objectives. When the quality objectives are established, how can you tell that you have met your objectives? The metrics will indicate to the management whether the product or service quality is improving. This requirement will enable management to evaluate current processes, set goals, and create the means for improvement opportunities.

 **TIPS:** Those metrics that relate to the product being certified must be defined in the supplier's quality objectives, and there must be plans in place for the collection, validation, and reporting of the data. One supplier had a five year plan with objectives for TL 9000 metrics that link to the business plan.

## 4.2 Quality System

> **4.2.2.C.1**
> *Lifecycle model*

 **Adder: 4.2.2.C.1: Lifecycle Model:** The supplier shall develop/implement a lifecycle model that will indicate various activities related to the entire development lifecycle from inception to the end of life.

## 22  Chapter Two

**Rationale:** In today's environment, the products/services are becoming complex. In multifaceted products the interdependencies and designs need to be considered up front to increase the serviceability, reliability, and availability. In order to ensure completeness of all activities and to make the product coherent, upfront identification of business related risks, project dependencies, costs, and resources will prevent major surprises in the end. This adder ensures that the supplier considers all activities during the life of the products and how they affect the overall quality.

**TIPS:** Utilization of a Gate Process will satisfy this requirement, but it must include the entire lifecycle, including operation, maintenance, and product disposal (when needed). It can't stop when the product is deployed and becomes generally available. Ongoing planning and support must be demonstrated. Also, it is critical that when a product/service is discontinued there be a planning and notification process that includes all customers.

If your scope only covers manufacturing, you can define your lifecycle to cover manufacturing only. If parts of the lifecycle are outsourced (contract manufacturing), the lifecycle must be linked into element 4.6, Purchasing. Gaining a companywide agreement on the interpretation and action plan involves more than expected time and resources. Do not underestimate the impact of this element. Involve Engineering and Sales early on with this task.

### 4.2.3 Quality Planning

**Adder: 4.2.3.C.1: Customer Involvement:** The supplier shall establish and maintain methods for soliciting and considering customer input for quality planning activities. Consideration should be given to establishing a joint customer-supplier quality improvement program.

**Rationale:** This clause addresses the value of input from the customer. Regular customer involvement ensures that customer's requirements are fulfilled related to product design, features, and domestic or international standards. The customer involvement may be in the selection of subcontractors, conducting joint reviews, or identifying requirements.

☞ **TIPS:** A joint customer-supplier quality improvement program fosters common understanding between the customer and the supplier concerning the product quality expectations and the manner in which deliverables (software, hardware, or service) are to be validated by the supplier. Without a joint understanding, a customer risks receiving a product that may not meet the functional or performance requirements.

Have periodic (at least twice yearly) meetings with individual customer(s) to share expectations, identify gaps, develop action plans, measure improvements, and review/update action register. Meet at the customer's site where new product has been introduced, within the first year, to discuss product specific issues.

An alternate method of communication is to hold periodic user group meetings.

This adder along with 4.21.2.C.2 Customer-Supplier Communication and 4.21.3.C.1 Customer Satisfaction caused the Marketing Department in one company to formally define and document a customer relationship process.

**Adder: 4.2.3.C.2: Long- and Short-term Planning:** The supplier's quality planning activities shall include long- and short-term plans with goals for improving quality and customer satisfaction. These plans shall address: (a) cycle time, (b) customer service, (c) training, (d) cost, (e) delivery commitments, and (f) product reliability.

**Rationale:** The planning enables the supplier to conduct risk assessments and identify mitigation activities, allocate appropriate resources, manage costs, and identify dependencies. Planning also forces the supplier to evaluate training needs, delivery commitments, process improvements, and potential corrective actions. Project planning helps suppliers ensure that schedules are met and target completion dates do not slip. Long term planning addresses maintenance activities, customer support, and installation. Note that "end of life" planning for the product is required in 4.4.2.C.3.

☞ **TIPS: C-NOTE A*** states that "Management with executive responsibilities should demonstrate active involvement in long- and short-term quality planning." Executives should not continually delegate their authority, but should show that they periodically attend quality-planning sessions. A lifecycle model should be defined and enforced for product development. Encourage planning in every major phase. Use tools like

---

*Notes are suggestions only not requirements.

Plan–Do–Check–Act (PDCA). Employee Goals and Objectives should cover items (b) through (e) and be tracked with employee reviews and appraisals. The supplier's long- and short-term objectives must address each of the items and the use of customer satisfaction data as an input. Quality improvements could be in the form of reduced cycle time, quicker response time, availability of training, reduced costs, on-time delivery, and product life.

One company uses a two-level approach. Executives develop high-level critical objectives. The operational level translates these into implementation activities. They also use Kaizen-type inputs from employees.

**Adder: 4.2.3.C.3: Subcontractor Input:** The supplier shall establish and maintain methods for soliciting and using subcontractor input for quality planning activities.

**Rationale:** Telecom suppliers are partnering more and more with contract manufacturers and other subcontractors. When a supplier purchases a product or service from a subcontractor, the supplier needs to clearly specify requirements and establish a procedure to collect information on quality related activities. Subcontractor feedback is related to practices, procedures, and activities that include product capabilities, design constraints, product validation requirements, or lessons learned. Communication between supplier and subcontractor should clearly establish the capability of the subcontractor to meet the supplier's requirements.

**TIPS:** In order to track the feedback, a formal method of collecting and documenting the feedback should be created by the supplier.

During development phase a project plan is written. Inputs are solicited from subcontractors for resources and commitments. Also they provide technical exchange of information and ideas. This is a partnering effort. Where subcontractors impact any of the quality planning items, there must be communications established with them.

**Adder: 4.2.3.C.4: Disaster Recovery:** The supplier shall establish and maintain methods for disaster recovery to ensure the ability to recreate and service the product throughout its lifecycle.

**Rationale:** Remember the San Francisco earthquake in October 1989 and the fire in a Con Edison substation at the foot of the Brooklyn Bridge. Communications were completely out for many hours after the earthquake and approximately 1000 companies were affected by the fire. A study by Contingency Planning Associates indicated that only 25 of the affected companies had any type of disaster recovery plan.

Disaster recovery is the most overlooked activity. With our growing reliance on the expanding technology, there certainly are increased risks when a disaster strikes. It is important to addresses disaster recovery planning by a supplier so that the product and service can be supported without major interruptions.

☞ **TIPS:** This adder requires that a recovery contingency plan be in place in the event that a customer's service fails.

For the supplier, a documented procedure should be written to provide back-up for software and storage off-site of the tapes or disks. Also, there is a need to provide back-up and support for key personnel whose loss would constitute a disaster. Natural disasters such as floods, tornadoes, and hurricanes have the ability to completely demolish a facility and these disasters should not be overlooked in terms of disaster recovery planning.

## 4.3 Contract Review

*4.3*
*Contract review notes*

There are no adders to the contract review process. However, there are two notes that are intended to make this process more consistent across the industry.

 **C-NOTE B:**\* The contract review process should include:

a. Product acceptance planning and review;

b. Handling of problems detected after product acceptance, including customer complaints and claims; and

c. Responsibility of removal and/or correction of nonconformities after applicable warranty period or during product maintenance contract period.

☞ **TIPS:** Item (a) is covered in detail by the following note, C-NOTE C. Item (b) requires details on how the customer notifies the supplier of a problem. Item (c) deals with how problems are handled after the product warranty period has elapsed.

---

\*Notes are suggestions only not requirements.

## 26 Chapter Two

**C-NOTE C:*** The product acceptance plan should include as appropriate:

a. Acceptance review process;

b. Acceptance criteria;

c. Documented test procedures;

d. Test environment;

e. Test cases;

f. Test data;

g. Resources involved;

h. Method(s) for problem tracking and resolution; and

i. Required acceptance test reports.

**TIPS:** Items (a), (b), and (c) require documentation be provided with these details. Item (d) requires any special environmental conditions, such as temperature control, heat tests, etc., to be documented. Items (e) and (f) require that data from previous tests be made available. Item (g) requires providing resource needs for acceptance tests. Item (h) requires showing how problems are resolved and tracked, such as mechanized diagnostics. Item (i) requires showing a sample of what an acceptance report will be.

### 4.4. Design Control

---

*Notes are suggestions only not requirements.

## Common Requirements

📠 **Adder: 4.4.1.C.1: Requirements Traceability:** The supplier shall establish and maintain a method to trace specified requirements through design and test.

🔍 **Rationale:** Often when a product or service progresses from one phase to the next, some of the initial requirements are lost along the way. The legal/statutory requirements, hazardous materials, or individual status of component parts necessitates the need for traceability. Effective traceability confirms that all requirements are present in the finished product and can be accounted for from initial phase through design and development to the finished product.

☞ **TIPS:** During the design and implementation process, requirements and design documents show mapping of design requirements through subsequent phases. An automated tool is used for creating the mapping.

Design reviews are performed at various stages to verify that the design process is not deviating from the specified requirements.

📠 **Adder: 4.4.2.C.1: Project Plan:** The supplier shall establish and maintain a project plan based upon the defined product life cycle model. The plan should include:

a. Project organizational structure;

b. Project roles and responsibilities;

c. Interfaces with internal and external organizations;

d. Means for scheduling, tracking, issue resolution, and reporting;

e. Budgets, staffing, and schedules associated with project activities;

f. Method(s), standards, documented procedure(s), and tools to be used;

g. References to related plans (for example, development, testing, configuration management, and quality);

h. Project-specific environment and physical resource considerations (for example, development, user documentation, testing, and operation);

i. Customer, user, and subcontractor involvement during the product lifecycle (for example, joint reviews, informal meetings, and approvals);

j. Management of project quality;

k. Risk management and contingency plans (for example, technical, cost, and schedules);

l. Performance, safety, security, and other critical requirements;

m. Project-specific training requirements;

n. Required certifications;

o. Proprietary, usage, ownership, warranty, and licensing rights; and

p. Post-project analysis.

**Rationale:** A project plan identifies major milestones of the entire project, from beginning to end, and also identifies individual work breakdown activities that must be completed to reach these milestones. It's important to control slippage in the schedule and identify and resolve issues before they impact the project adversely. The project plan defines resources and schedules, identifies risks, indicates activities assigned to appropriate resources, and highlights dependencies. Individual functional groups within a project may have their own mini-project plan which then links into an overall master project plan.

Development plan, test plan, configuration management plan, support plan, risk identification/mitigation plan, and security plan may be part of the project plan or separate documents within the project.

**TIPS:** Remember, this is a *should* requirement and may be replaced by an equivalent process that meets the intent of the requirement.

Use two levels in project plans. The highest level covers all current activities. For each activity a template can be used to cover the key points.

Items (a) through (e) require the supplier to show that there are adequate human resources planned for the project. Item (f), (g) and (h) require the supplier to show that there are adequate documents and tools planned to support the project. Item (i) requires the supplier to show that there has been involvement outside the design group, such as customer, user, and subcontractor inputs. Item (j) requires the supplier to show the project quality plan. Item (k) requires the supplier to show what contingency will be used if the project resources require change. Item (l) requires that the supplier address these specific issues within their requirements. Item (m) requires that the supplier provide training for the project. Item (n) requires the supplier to identify any special user certifications needed. Item (o) requires the supplier to identify all legal aspects of the project. Item (p) requires the supplier to review the results of the finished project and analyze the results. In the project plan, identify each major development area, assign personnel, and plan deliverables, schedules, and resources.

Track resources and schedules during development. Use the gate lifecycle model to establish project controls.

**C-NOTE E:**[*] The project plan and related plans may be an independent document, a part of another document, or composed of several documents.

**C-NOTE F:**[*] General work instructions defining tasks and responsibilities common to all development projects need not be replicated as part of a documented project plan.

These notes provide the supplier the latitude to reference the same document(s) for several different projects.

**Adder: 4.4.2.C.2: Test Planning:** Test plans shall be documented and results recorded. Test plans should include:

a. Scope of testing (for example, unit, feature, integration, system, acceptance);

b. Types of tests to be performed (for example, functional, boundary, usability, performance, regression, interoperability);

c. Traceability to requirements;

d. Test environment (for example, relevancy to customer environment, operational use);

e. Test coverage;

f. Expected results;

g. Data definition and database requirements;

h. Sets of tests, test cases (inputs, outputs, test criteria), and documented test procedure(s); and

i. use of external testing.

**Rationale:** Test planning activities ensure that appropriate robust tests are identified. The planning takes into consideration interdependencies of various modules and related tests, identifies problem-reporting procedures, and enforces implementation of time frames for problem resolution depending on the severity of the problem. The test plan is a "blue-print" of the testing activities and identifies all the inputs and outputs required to test the thoroughness of the product.

---

[*]Notes are suggestions only not requirements.

## 30  Chapter Two

☞ **TIPS:** Remember, this is a *should* requirement and may be replaced by an equivalent process that meets the intent of the requirement.

For each test phase have a test specification and test plan. Test specifications contain details of tests including test scenarios. Test plans contain scheduling and resource commitments.

Items (a) to (c) require that the supplier document the types and scopes of testing and how they relate to the project requirements. Item (d) requires that the supplier document the project impact on customer environment and use. Items (e) and (f) require that the supplier document test coverage and results. Item (g) requires that the supplier identify specific software items to be included in the database. Items (h) and (i) require that the supplier document test procedures and their results.

📓 **C-NOTE G*** states that testing may be covered at several levels. This note is intended to point out that test planning should be carried out for various levels including algorithms, modules, sub-systems, etc.

📠 **Adder: 4.4.2.C.3: End of Life Planning:** The supplier shall establish and maintain a documented procedure(s) for the discontinuance of manufacturing and/or support of a product by the operation and service organizations. This documented procedure(s) should include:

a. Cessation of full or partial support after a certain period of time;

b. Archiving of product documentation and software;

c. Responsibility for any future residual support issues;

d. Transition to new product, if applicable; and

e. Accessibility of archived copies of data.

🖋 **Rationale:** The End of Life planning ensures common understanding between marketing, development, and customer support regarding when the product/service will reach end of life. This planning also assists organizations in notifying customers regarding when the support of a particular product will be stopped. This helps in managing customer expectations.

☞ **TIPS:** Remember, this is a *should* requirement and may be replaced by an equivalent process that meets the intent of the requirement.

Item (a) requires that the supplier notify their customers when a product's support is going to change. Items (b) and (e) require that the supplier retain discontinued product documentation and software and that it is accessible. Item (c) requires that the supplier plan for support of residual

---

*Notes are suggestions only not requirements.

problems. Item (d) requires that the supplier inform the customers how to transition to new product.

In one company, the Business Management department handles end of life planning.

**4.4.3 Organizational and Technical Interfaces. C-NOTE H:*** The supplier should establish communication methods for dissemination of product requirements, and changes to requirements to all impacted parties identified in the project plan.

**TIPS:** This note suggests that the supplier should have a process in place to deliver project changes and a listing of all personnel requiring the information.

**Adder: 4.4.4.C.1: Customer and Subcontractor Input:** The supplier shall establish and maintain methods for soliciting and using customer and subcontractor input during the development of new or revised product requirements.

**Rationale:** This section of the standard emphasizes establishment of a formal process by the supplier to collect and implement input from customer and subcontractor. The input may be related to the development methodology, new features, or changes to the existing product. A structured process of collecting and addressing the input ensures that all customer/subcontractor concerns and ideas are documented and addressed.

**TIPS:** This adder requires that the supplier contact customers and subcontractors for their inputs on new or revised products. For details, see 4.2.3.C3 for subcontractor input and 4.2.3.C.1 for customer input.

**Adder: 4.4.4.C.2: Design Requirements:** Design requirements shall be defined and documented and should include:

a. Quality and reliability requirements;

b. Functions and capabilities of product;

c. Business, organizational, and user requirements;

d. Safety, environmental, and security requirements;

e. Installability, usability, and maintainability requirements;

f. Design constraints; and

g. Testing requirements.

---

*Notes are suggestions only not requirements.

**32  Chapter Two**

**Rationale:** This guarantees that design requirements such as reliability, availability, and serviceability are taken into consideration in the early stages of development. Safety as well as regulatory requirements are also considered at the design stage. Design limitations must be identified. Documented design requirements facilitate ease in implementing future changes.

**TIPS:** Depending upon the type of product/service, the following requirements documents may be needed. They can be addressed in a single document:

- High-level requirements;
- Design requirements;
- Architectural requirements;
- Testing requirements;
- Installation;
- User documentation;
- Performance requirements;
- Packaging requirements;
- Tolerance requirements;
- Safety requirements;
- Security requirements;
- Maintainability requirements.

The following list is relative to installation services only.

a. Service requirements, such as crimped connections, soldering, torque, appearance, etc.;

b. Service capabilities, such as 24/7 support, training output, and technical support;

c. User requirements, such as installation interval, personnel, tools, training, etc.;

d. Safety requirements, such as electrical, chemical, personal, etc.;

e. Usability requirements, such as product or service affected;

f. Constraint requirements, such as service protection, tooling, environmental, etc.;

g. Testing requirements, such as test sets, instructions, etc.

**4.4.7 Design Verification. C-NOTE I:**[*] See 4.10 for related requirements.

**TIPS:** This note points out the parallel between design verification and inspection and testing of products.

**4.4.8 Design Validation. C-NOTE K:**[*] It may be beneficial to include customers or a third party during various validation stages.

**TIPS:** This note points out a means of improving communication with the customer and within the supply chain.

**Adder: 4.4.9.C.1: Change Management Process:** The supplier shall establish and maintain a process to ensure that all requirements and design changes, which may arise at any time during the product lifecycle, are managed in a systematic and timely manner and do not adversely affect quality and reliability. Management of the changes should include:

a. Impact analysis;
b. Planning;
c. Implementation;
d. Testing;
e. Documentation;
f. Communication; and
g. Review and approval.

**Rationale:** No matter how well the original requirements are collected, documented, and reviewed, the changes are unavoidable. Some causes are technology transition, rapid growth, changing business needs, constantly changing customer needs, and competition. This adder ensures that the supplier establish a process in which all the changes to a project's contractual items are documented and managed.

**TIPS:** Corrective action procedure tracks all changes. For hardware, use the product change notice procedures to document and track changes. There are different levels of changes such as class A, D, etc.[4]

Changes to service requirements must be planned and implemented, using a systematic method, such as first office application or laboratory trials. The impact of the change should be documented and provided to users of the service.

---

[*]Notes are suggestions only not requirements.

One process for change management consists of three steps:

1. Initiate change request: A change can be requested by any staff member of the project, by the customer, or by the subcontractor.

2. Change Request Review: All changes are reviewed to see overall project impact in regards to resources, cost, and schedule. If the impact is low, the change is implemented. If the impact is high, detailed review is conducted to scope the impact.

3. Implementation of approved changes: If approved, a requested change is implemented. This may require changing the hardware, software, service, and related documentation or processes.

Change management entails keeping track of changes by documenting the date when change was initiated, description, reason, and impact of a change. The process of change management may be handled by the maintenance organization, which assures that everyone involved is notified of any changes and keeps abreast of associated risks and dependencies.

**Adder: 4.4.9.C.2: Informing Customers:** The supplier shall establish and maintain a documented procedure(s) to ensure that customers are informed when design changes affect contractual commitments.

**Rationale:** The contract is a binding document between the supplier and the customer. The supplier must have an established, documented process to notify the customer of the changes that may affect the contractual conditions. Notifying the customers of any adverse affects on the contractual obligations will help manage the customer expectations and customer satisfaction.

**TIPS:** You may use the Telcordia-recommended procedure for product change notices to satisfy this requirement.[4]

Changes to service methods performed by customers, such as self-install, must include impact on intervals, tools, training, etc., and be provided to customers using service.

## 4.6 Purchasing

**C-NOTE N:**[*] The reference to "(see 3.1)" above references ANSI/ISO/ASQC Q9001–1994, which is not included in this document.

---

*Notes are suggestions only not requirements.

☞ **TIPS:** This note defines product as the "result of activities or processes, [that] may include service, hardware, software, processed material or a combination thereof. A product can be tangible (e.g. assemblies or processed materials) or intangible (e.g. knowledge or concepts) or a combination thereof. For the purposes of this international standard, the term 'product' applies to the intended product offering only and not to unintended 'by-products' affecting the environment. This differs from the definition given in ISO 8402."[1]

**4.6.1.C.1 Purchasing Procedure(s)**

```
4.6.1.C.1
Purchasing procedure(s)
```

The documented purchasing procedure(s) shall include:

a. Product requirements definition;

b. Risk analysis and management;

c. Qualification criteria;

d. Contract definition;

e. Satisfaction of proprietary, usage, ownership, warranty, and licensing rights;

f. Planning for future support of the product;

g. Ongoing supply-base management and monitoring;

h. Subcontractor Selection criteria;

i. Subcontractor re-evaluation;

j. Feedback to key subcontractors based on data analysis of subcontractor performance.

✎ **Rationale:** There has been a large increase in the use of outsourcing of components, assemblies, and services in recent years. Because of this, it is necessary to specify requirements which cover problem areas identified in the past.

**TIPS:** Items (a) through (f) must be covered in a contract. Items (g) through (j) require the purchasing organization to gather ongoing subcontractor evaluation data that tracks subcontractor's historical performance, such as tools, test sets, product installability, documentation, training, installation work, etc., and provide feedback to subcontractor. Item (b), risk analysis and management, is especially crucial because of the pressures today on the supply chain. Often many competitors use the same suppliers, which may affect their ability to meet schedules.

**C-NOTE O:**[*] This documented procedure(s) should be applicable to off-the-shelf product. This typically includes original equipment manufacturer (OEM) products used in manufacturing and commercial off-the-shelf (COTS) products used in software systems.

## 4.10 Inspection and Testing

> 4.10
> Inspection and
> testing note

**C-NOTE Q:**[*] See 4.4.7 and 4.4.8 for related requirements.

**TIPS:** This note points out the parallel between design verification and inspection and testing of products.

## 4.13 Control of Nonconforming Product

> 4.13.2.C.1
> Trend analysis

**Adder: 4.13.2.C.1: Trend Analysis:** Trend analysis of discrepancies found in nonconforming product shall be performed on a defined/regular basis and results utilized as input for corrective and preventive action.

---

[*]Notes are suggestions only not requirements.

*Rationale:* Controlling nonconforming products results in continual improvement. The history of nonconforming products must be kept and the data must be utilized to conduct root cause analysis (RCA). The results of the RCA activities are used to implement processes that would correct the nonconformance and also prevent it from occurring in the future. Trend analysis gives a barometer of the implemented processes to see if these processes are yielding desired quality results.

*TIPS:* This adder requires that trend analysis be performed on discrepancies at regular intervals and fed to the preventive/corrective action process.

Use documented procedures for doing measurement analysis. Standardized reports are reviewed on a regular basis for corrective action.

One organization did not have a database for tracking corrective actions. A process was put in place to track the corrective actions on the ISO 9000 database.

## 4.14 Corrective and Preventive Action

> **4.14**
> **Corrective and preventive action notes**

*C-NOTE S:** Consideration should be given to include training as part of implementing corrective and preventive actions.

*TIPS:* This note suggests that training could be part of the preventive/corrective action solution.

*C-NOTE T:** Negative deviations from plans are considered nonconformances.

*TIPS:* This note is meant as an aid in determining if a condition is considered a nonconformance.

*C-NOTE U:** Effective corrective action ensures that additional problems have not been introduced.

---

*Notes are suggestions only not requirements.

☞ **TIPS:** This note suggests that a review of corrective action implementation should include assuring that new problems have not been introduced.

## 4.15 Handling, Storage, Packaging, Preservation, and Delivery

```
┌─────────────┐ ┌─────────────┐
│  4.15.1.C.1 │ │  4.15.1.C.2 │
│  Work areas │ │  Anti-static│
│             │ │  protection │
└─────────────┘ └─────────────┘
```

📇 **Adder: 4.15.1.C.1: Work Areas:** Areas used for handling, storage, and packaging of products shall be clean, safe, and organized to ensure that they do not adversely affect quality or personnel performance.

✎ **Rationale:** Working environment contributes to the overall quality of the products as well as affects the morale and well-being of the personnel. Poor working conditions may have adverse impact on the product. External factors such as handling, storage, and packaging of products can contribute to the nonconformity of the product. Nonconforming, returned products must be organized and stored adequately to eliminate erroneously sending it back to a customer.

☞ **TIPS:** This adder enforces the need for clean, safe, and organized work areas. Training of personnel involved with development and manufacturing may be necessary. Where "clean rooms" are involved, special environments must be maintained and special procedures followed.

📇 **Adder: 4.15.1.C.2: Anti-Static Protection:** Anti-static protection shall be employed where applicable, for components and products susceptible to electrostatic discharge (ESD) damage. Consider such components and products as: integrated circuits, printed wiring board assemblies, magnetic tapes and/or disks, and other media used for software or data storage.

✎ **Rationale:** Equipment such as microprocessors, printed board assemblies, and magnetic storage media can be damaged by ESD. Taking precautions will prevent damage in the engineering lab, the manufacturing or distribution centers, or at the customer site.

☞ **TIPS:** Establish a General Operating Procedure for ensuring that protection is provided while handling ESD-sensitive components, assemblies, and equipment. The main tools of protection are wrist straps, floor mats, and anti-static wrapping materials. Training of appropriate personnel is required by 4.18.C.4.

## 4.18 Training

**Adder: 4.18.C.1: Course Development:** The supplier shall establish and maintain a process for planning, developing and implementing internally developed training courses.

**Rationale:** For any type of training, planning is required. Some subjects to be considered in the training plans include:

- experience of audience in the subject matter
- contents of the training
- objectives of the training
- length of the training
- schedule
- training equipment
- training facilities
- piloted training and incorporation of the feedback to continually improve the contents of the training
- establishing a formal process to keep record of attendees who successfully complete the training
- determining a method of reviewing the effectiveness of the training

If in-house developed training has no structured process, it will become obsolete in a short period of time.

☞ **TIPS:** Training departments should have a procedure on developing courses and maintaining records. The procedures should include a means of reviewing the effectiveness of the training. This may consist of on-the-job testing, course surveys of students, interviewing the student's managers, and review of process and product defects. A training curriculum can help establish "centers of excellence" or an expert network.

📇 **Adder: 4.18.C.2: Quality Improvement Concepts:** Those employees that have a direct impact on the quality of the product, including management with executive responsibility, shall be trained in the fundamental concepts of quality improvement, problem solving, and customer satisfaction.

🔍 **Rationale:** Most employees in an organization have a direct impact on the quality of the product. It is especially important for management with executive responsibility to have quality awareness and understand the concepts of total quality management that will result in continual improvement of the processes and products. Understanding of quality concepts by all employees will affect the robustness of the products, and ultimately customer satisfaction.

☞ **TIPS:** This adder requires that quality improvement concepts be part of most of the employee's training curriculum, including executives.

In one organization, a new employee has to have an overview of quality concepts, and every quarter a schedule goes to each employee on available courses.

The overview course was given to supervisors, engineers and management. Supervisors cascaded the training to the shop personnel. Senior leadership can demonstrate commitment by attending class with their staff.

📇 **Adder: 4.18.C.3: Training Requirements and Awareness:** Training requirements shall be defined for all positions that have direct impact on the quality of the product. Employees shall be made aware of training opportunities.

🔍 **Rationale:** Behind the success of every company is the experience and knowledge of the individuals that are part of the company. Depending on the job to be performed, each position within a company requires special skills. Identification of the skill-set needed to accomplish a particular job and providing adequate training assists the employees in fulfilling their job related responsibilities. Well-trained employees are the most valuable asset for the success of an organization.

☞ **TIPS:** This adder requires that there be training requirements specified for each job position, and that these requirements be reviewed with employees.

**Adder: 4.18.C.4: ESD Training:** All employees with functions that involve any handling, storage, packaging, preservation, or delivery of ESD-resistive products shall receive training in electrostatic discharge (ESD) protection prior to performing their job.

**Rationale:** As mentioned in 4.15.1.C.2, ESD damages components and products such as integrated circuits, printed wiring board assemblies, magnetic tapes and/or disks, and other media used for software or data storage. In order to prevent this damage, employees should be provided with training so they will be aware of the ramifications of ESD and take precautionary measures to avoid damage.

**TIPS:** This adder requires ESD training prior to working on ESD sensitive items.

**Adder: 4.18.C.5: Advanced Quality Training:** The supplier shall offer training in statistical techniques, process capability, statistical sampling, data collection and analysis, problem identification, problem analysis, and corrective and preventive action, as appropriate.

**Rationale:** Every employee should be familiar with Deming's "Plan–Do–Check–Act" methodology:

- Planning for a new process
- Implementing the new process
- Collecting information and a statistical sampling of data to make sure the new process works
- Analyzing and planning a new approach if the sampling of data reflects no improvement

Product quality is affected by the skills and knowledge of the personnel.

**TIPS:** Training in process improvements, verification, and validation activities, will have a direct impact on the quality of the products. Root cause analysis is a vital activity in identifying the causes of the problems, and the skill-set of the employees to address the causes will eliminate similar problems in the future.

This adder requires that training for advanced quality techniques be available, as appropriate. This does not mean the supplier must develop these courses internally, but they must have a way of training employees in these techniques. Commercial training courses may be used.

**Adder: 4.18.C.6: Training Content:** Where hazardous conditions exist, training content should include the following: (a) task execution,

(b) personal safety, (c) awareness of hazardous environment, and (d) equipment protection.

*Rationale:* Training should be given in order to eliminate and reduce employee injuries or damage to the business and related equipment. The training should specifically address:

- personal safety to avoid injuries
- the process to execute the job
- awareness of hazardous environments that might damage the product, equipment, and personnel

*TIPS:* This adder requires that there be training courses available for personal safety, hazardous material handling, and service protection.

Every new employee should go through training on hazardous materials.

## 4.19 Servicing

*Adder: 4.19.C.1: Supplier's Support Program:* The supplier's quality program shall ensure that customers are provided support to resolve product related problems.

*Rationale:* To ensure customer satisfaction, it is important for the supplier to designate resources to assist the customer in product-related problems or questions. The support personnel shall be easily available to ensure that any inquiry regarding a product, customer concern, or problem is satisfactorily addressed. The support program will ensure that a quality plan is established to resolve product-related problems within an acceptable timeframe. If the customer and supplier are in two different time zones, consideration should be given by the supplier to the support hours in order to provide the support during mutually acceptable timeframes.

*TIPS:* Some suggestions are: (1) a 24/7 call center, (2) a web site with "frequently asked questions," and (3) a chat room for individual products.

This requirement may be handled by customer and product support organizations, which provide expertise on an as-needed basis. This adder requires technical support be available to fix customer services complaints.

**Adder: 4.19.C.2: Service Resources:** The supplier shall provide customer contact employees with appropriate tools, training, and resources necessary to provide effective and timely customer service.

**Rationale:** To ensure that the customer is effectively supported, the experience, knowledge, and skill-set of customer support individuals at the supplier site is essential. Training, tools, and documented processes will equip the support staff to do their job in a timely and efficient manner.

The customer may need support in:

- Installation problems
- Product related issues
- Delivery related questions
- Invoice related questions

**TIPS:** This adder is applicable to maintenance and field support-type functions, not traditional central office installations. Special courses are developed and given to customer contact personnel.

**Adder: 4.19.C.3: Notification About Problems:** The supplier shall establish and maintain documented procedure(s) to notify all customers who may be affected by a reported problem that is service-affecting.

**Rationale:** Early notification to the customers of any service-affecting problems will assist in reduced customer calls and will keep the customers abreast of quality related issues with the product. The proactiveness on the supplier's part builds customer confidence.

**TIPS:** This adder requires that there be a customer notification process for reporting service-affecting problems.

Document a procedure on how to contact the customers when a problem occurs. This will assist the supplier in managing customer anxiety and frustration.

**Adder: 4.19.C.4: Problem Severity:** The customer and supplier shall jointly assign severity levels to customer reported problems based on the impact to the customer. The severity level shall be used in determining the timeliness of the supplier's response.

**Rationale:** A problem that may be considered critical by a customer may not be necessarily perceived with a similar criticality by the supplier. Having the supplier and customer jointly identify and understand the

## 44 Chapter Two

criticality of the problem will assist in eliminating the wait time frustration for the customer. Understanding how a defect would affect the customer's ability to provide service will allow the supplier to allocate adequate resources to resolve the defect in an efficient manner.

**TIPS:** This adder requires that there be customer input in determining the severity of a reported problem and that the severity be based upon customer impact.

At one company, Technical Assistance Center employees receive customer complaints and assign severity per customer suggestions. At times subject matter experts review the severity assigned and negotiate a change with the customer.

**Adder: 4.19.C.5: Problem Escalation:** The supplier shall establish and maintain documented escalation procedure(s) to resolve customer reported problems.

**Rationale:** When a problem cannot be resolved by the first-level customer-support personnel, it is important for the supplier to have a documented procedure on how the problem will be escalated to a second-level support organization so it can be resolved in a timely fashion. In many companies, there are multiple levels of support organizations and a problem may be transferred from one level to the next when it cannot be resolved by the initial level of support.

**TIPS:** Have the appropriate support people on call.

In one organization, problems are documented at the customer assistance center. They try to resolve them online. If they can't solve them, they are escalated to appropriate organizations.

## 4.20 Statistical Techniques

> 4.20.1.C.1
> Process measurement

**Adder: 4.20.1.C.1: Process Measurement:** Process measurement shall be developed, documented, and monitored at appropriate points to ensure continued suitability and promote increased effectiveness of processes.

**Rationale:** The relationship between the quality of a process and the quality of the product is multifaceted. This relationship depends on many factors, such as robustness of the processes, employees who follow the processes, and tools that are used to accomplish the processes. When the product quality deteriorates, it is difficult to determine which one of the three factors was the biggest contributor. For this reason, process measurements shall be kept to give accurate and meaningful information. The measurements give meaningful data for improving the reliability of the product and improving the processes used to develop the product. The purpose of the measurements is to determine tangible data for managing, estimating, and sizing projects/products.

**TIPS:** This adder requires that there be process measurements taken and monitored to ensure process effectiveness.

## 4.21 Quality Improvement and Customer Satisfaction

**Adder: 4.21.1.C.1: Quality Improvement Program:** The supplier shall establish and maintain a documented Quality Improvement Program to improve (a) customer satisfaction, (b) quality and reliability of the product, and (c) other processes/product/services used with the company.

**Rationale:** All the above mentioned adders are related to the total quality concepts. The quality elements such as measurements, data analysis, feedback to supplier related to the product, documented procedures, employee training of the procedures, customer satisfaction survey, and quality records are the components of sound business practices. Evaluating the results of the quality improvement program in periodic intervals shows whether the program is bringing desired results. If the results are sagging, this is an indication to review the entire quality improvement program so effective changes can be implemented.

☞ **TIPS:** This adder requires that there be a documented quality improvement program.

📇 **C-NOTE V:*** Inputs to the continuous improvement process may include lessons learned from past experience, lessons learned from previous projects, analysis of metrics and post-project reviews, and comparisons with industry best practices.

This note suggests using lessons learned from past experience.

The supplier needs methods of gathering and validating data. The difficulty in using customer satisfaction data is translating the inputs into actions. This can be done by linking this data into the corrective and preventive action process.

🖩 **Adder: 4.21.1.C.2: Employee Participation:** The supplier shall have methods encouraging employee participation in the quality improvement process.

🔍 **Rationale:** The livelihood of any company depends on the employees' knowledge and experience and the quality of the products. Employees of the organization have a direct impact on the way a product is built and serviced. Their participation in identifying the problem areas of the product and processes will result in root cause analysis and preventive as well as corrective actions. Encouraging employees to voice concerns regarding deviations from pre-established processes or making them "owners" of quality of the product can result in better products and improved employee morale.

☞ **TIPS:** This adder requires that employees have an opportunity to participate in the QI process.

One company uses a Kaizen program and monitors the number of suggestions submitted. Another method is the use of quality circles. Reports and charts can be published on customer feedback.

🖩 **Adder: 4.21.1.C.3: Supplier Performance Feedback:** The supplier shall inform employees of its quality performance and the level of customer satisfaction.

🔍 **Rationale:** When employees know about customer satisfaction and quality performance results they feel a part of the business. Since their work has a direct relationship to quality performance, raising their awareness and participation will result in overall improvements.

☞ **TIPS:** In many companies the employee satisfaction numbers are based on a rigorous statistical analysis, and the satisfaction number is tied

---

*Notes are suggestions only not requirements.

to employee bonuses and other compensation. In these companies, trend analysis of customer satisfaction is kept and each year the "bar" for the satisfaction is raised from the previous year results. If the company reaches the pre-established satisfaction number, the staff is rewarded with bonuses or other high compensation. Many companies use continuous customer satisfaction surveys in lieu of a yearly survey.

One company displays customer report cards on its web site based on CSQP$^{SM}$.[5,6] They are also displayed throughout the organization.

**Adder: 4.21.2.C.1: Management Commitment:** Management with executive responsibility shall demonstrate active involvement in establishing and maintaining customer-supplier relationships.

**Rationale:** Management has the authority to make changes in an organization's pre-established processes if the process does not bring desired results. Similarly, if there are any issues with the customer-supplier relationship, a member from senior management has the ability to address the issues and manage the relationship with the customer in a timely fashion.

**TIPS:** This adder requires that supplier executives communicate directly with the customer. It can't be delegated. This may include customer visits, attendance at user group meetings, or being active in the QuEST Forum.

**Adder: 4.21.2.C.2: Customer-Supplier Communication:** The supplier shall establish and maintain a documented procedure(s) for communicating with selected customers. The documented procedure(s) shall include: (a) a strategy and criteria for customer selection; (b) a method for sharing customer and supplier expectations and improving the quality of products; and (c) a joint review with the customer at defined intervals covering the status of customer-supplier shared expectations and including a method to track the resolution of issues.

**Rationale:** The telecom industry has a history of strong customer-supplier communication. This is because of the complexity of the products and the need to continually add features.

**TIPS:** Many companies have documented procedures determining the customers' quality expectations. The procedures may include joint reviews. The feedback from the reviews is valuable in influencing new products or service, adding new features, and improving overall quality of the products.

Joint reviews at regular intervals will assist in uncovering defects or issues on an ongoing basis. This will assist in project management and will prevent "surprises" from showing up at the end! The procedure in CSQP$^{SM}$ is to hold semi-annual meetings with key customers. The agenda includes strategies and change in product line.

**C-NOTE W:**[*] It is recognized that it is not possible for the supplier to provide the same level of communications with all its customers. The level provided may depend upon the amount of business with the customer, the history of problems, customer expectations, and other factors. See Appendix F of Reference 2, "Guidance for Customer-Supplier Communications."

This note provides guidance on how to choose selected customers for joint reviews.

**Adder 4.21.3.C.1: Customer Satisfaction:** The supplier shall establish and maintain a method to collect data directly from customers concerning their satisfaction with provided products. The supplier shall also collect customer data on how well the supplier meets commitments and its responsiveness to customer feedback and needs. This data shall be collected and analyzed, and trends shall be kept.

**Rationale:** Customer satisfaction is a key to the long-term survival of any business. Monitoring customer satisfaction for the quality and support of the products will enable the supplier to take immediate corrective actions in the event that customer satisfaction starts slipping. Delay in responding to customer questions, issues, or problems can affect the customer satisfaction, resulting in loss of business and customers.

**TIPS:** Adverse impacts should be monitored and managed. The turn-around time to resolve the problems or to get back to the customer with a quick response is important.

**Adder: 4.21.4.C.1: New Product Introduction (NPI):** The supplier shall establish and maintain documented procedure(s) for introducing new products.

**Rationale:** The intent of this process is to provide a vehicle for launching a new product so it can be supported in a timely fashion. Various logistics of the product shall be considered prior to releasing it. These logistics may include customer support readiness, sales readiness, manufacturing readiness, and service readiness.

**TIPS:** With complex products entering the networks, it is important to have a strong product support organization ready for product launches.

---

[*]Notes are suggestions only not requirements.

# END NOTES

1. ANSI/ISO/ASQC Q9001–1994, *Quality Systems—Model for Quality Assurance in Design/Development, Production, Installation and Servicing*, Second Edition (Milwaukee, WI: American Society for Quality Control, 1994).
2. *TL 9000 Quality System Requirements*, Book One, Release 2.5 (The QuEST Forum, 1999).
3. Alka Jarvis and Vern Crandall. *Inroads To Software Quality* (Englewood Cliffs, NJ: Prentice Hall, 1997).
4. GR-209-CORE, *Generic Requirements for Product Change Notices* (Morristown, NJ: Telcordia Technologies, Issue 3, September 1998).
5. GR-1202-CORE, *Generic Requirements for Customer Sensitive Quality Infrastructure* (Morristown, NJ: Telcordia Technologies, Issue 1, October 1995).
6. SR-3535, *Bellcore* CSQP$^{SM}$ *Program* (Morristown, NJ: Telcordia Technologies, Issue 1, October 1995).

# 3

# Hardware, Software, and Service Requirements

## INTRODUCTION

As mentioned in the preface, the TL 9000 requirements are based on ANSI/ISO/ASQC Q9001–1994.[1] Of the 83 "adders," 46 are hardware, software, service, or a combination of 2 types. In this chapter we repeat the requirements from the requirements handbook[*,2] give a rationale for the requirement, and provide "tips" from organizations that have implemented TL 9000. The reader should focus on the tips because they provide real implementation information.

The standard definition for the word *shall* in ANSI/ISO/ASQC Q9001–1994 is used in TL 9000, but *should* has a different meaning. The word *shall* is used for mandatory requirements that must be adhered to; *should* on the other hand indicates preferred practices. Suppliers choosing other approaches must be able to demonstrate that their approach meets the intent of TL 9000.

## CODES USED TO IDENTIFY THE TYPE OF REQUIREMENT

Individual codes were given to identify each adder as hardware, software, service specific, or when the adder is common to hardware, service, or software.

C: Common (hardware, software, and service)

HS: Hardware and software

---

*The adders and notes in chapters 2 and 3 are repeated verbatim with permission of the American Society for Quality, publisher of *TL 9000 Quality System Requirements* for the QuEST Forum.

**Table 3.1** Adders.

| ANSI/ISO/ASQC Q9001–1994 Element | Software | Hardware | Service | Hardware & service | Hardware & software |
|---|---|---|---|---|---|
| 4.1 | | | | | |
| 4.2 | 1 | | | | |
| 4.3 | | | | | |
| 4.4 | 7 | 6 | 2 | | |
| 4.5 | 1 | | | | |
| 4.6 | | | | | |
| 4.7 | | | | | |
| 4.8 | | 2 | | | 2 |
| 4.9 | 2 | 1 | 2 | 3 | |
| 4.10 | 1 | 2 | | 2 | |
| 4.11 | | 1 | | | |
| 4.12 | | | | | |
| 4.13 | | | | | |
| 4.14 | | | | | |
| 4.15 | 2 | 1 | | | |
| 4.16 | | | | | |
| 4.17 | | | | | |
| 4.18 | | | | | |
| 4.19 | 2 | 1 | | | 3 |
| 4.20 | | | | | |
| 4.21 | | 1 | 1 | | |
| Total | 16 | 15 | 5 | 5 | 5 |

HV: Hardware and service

H: Hardware only

S: Software only

V: Service only

Table 3.1 identifies all the adders and corresponding ANSI/ISO/ASQC Q9001–1994 elements and indicates when the adder applies to a specific area such as software, hardware, service, or combinations of two.

## HARDWARE/SOFTWARE ADDERS

TL 9000 has identified adders that are common to both software and hardware design and development in Sections 4.8 Product Delivery and 4.19

Servicing. We identify these adders in the next few pages, give you a rationale as to why they are a value-add to a quality system and provide tips on implementing them.

## 4.8 Product Delivery

```
┌─────────────────────┐  ┌─────────────────────┐
│      4.8.HS.1       │  │      4.8.HS.2       │
│   Configuration     │  │       Product       │
│  management plan    │  │    identification   │
└─────────────────────┘  └─────────────────────┘
```

**Adder: 4.8.HS.1: Configuration Management Plan:** The supplier shall establish and maintain a configuration management plan which should include: (a) identification and scope of the configuration management activities; (b) schedule for performing these activities; (c) configuration management tools; (d) configuration management methods and documented procedure(s); (e) organizations and responsibilities assigned to them; (f) level for required control for each configuration time; and (g) point at which items are brought under configuration management.

**Rationale:** Configuration management (CM) controls information describing the physical and logical characteristics of resources, and the relationships between those resources. Well-defined CM is a key component to the delivery of any product or service and assists in identifying version number, part number, or documentation. CM gives important information such as the versions of products supported in the field and the versions of product that an individual customer may have. There are several automated tools in the industry to assist in the CM activity.

**TIPS:** A documented CM process will identify resources that are responsible for the CM activity, description of when a particular object of software or hardware would come under the CM, the tools used for CM, and a schedule and list of all deliverable/activities that would be covered under CM. For software the process should cover builds, sanity checks, and testing procedures.

**HS-NOTE P:**[*] General work instructions defining general configuration management tasks and responsibilities need not be replicated as part of a specific documented configuration management plan.

**Adder: 4.8.HS.2: Product Identification:** The supplier shall establish and maintain a process for the identification of each product and the level of required control. For each product and its versions, the following shall be identified as appropriate: (a) documentation; (b) associated tools needed for product re-creation; (c) interfaces to other software and hardware; and (d) software and hardware environment.

**Rationale:** Design documents and architectural drawings are extremely important when a software or hardware product has to be re-created. In case of a component failure, product identification will make it easier to replace the component. The traceability of product and versions also allows easy tracking of nonconforming hardware or software and facilitates the root cause analysis. When a product is recalled due to severe defects, it is usually recalled by identifying part number. A missing-part identification will create a massive problem at recall time or at the warranty fulfillment time.

**TIPS:** Coordinate the hardware aspects of this requirement with the hardware traceability requirements (4.8.H.1 and 4.8.H.2).

## 4.19 Servicing

| 4.19.HS.1 Emergency service | 4.19.HS.2 Problem resolution configuration management | 4.19.HS.3 Installation plan |
|---|---|---|

**Adder: 4.19.HS.1: Emergency Service:** The supplier shall ensure that services and resources are available to support recovery from emergency failures of product in the field throughout its expected life.

**Rationale:** Emergency service is a key factor in creating customer confidence and satisfaction. The supplier needs to allocate adequate resources throughout the product lifecycle to assist the customer in case of emergencies.

**TIPS:** In one company the repair services organization has a procedure for supporting the customer based on the Telcordia CSQP$^{SM\ 3,4}$ procedures.

**Adder: 4.19.HS.2: Problem Resolution Configuration Management:** The supplier shall establish an interface between problem resolution and configuration management to ensure that fixes to problems are incorporated in future revisions.

**Rationale:** This activity ensures that fixes in the current product are also incorporated in the future revisions so the same errors do not plague the new revisions.

☞ **TIPS:** In one company, configuration management tools and problem reporting tools are integrated. Review boards assure that problem solutions are carried forward in future versions.

🖫 **Adder: 4.19.HS.3: Installation Plan:** The supplier shall establish and maintain a documented installation plan. The installation plan shall identify resources, information, and installation events. Installation events and results shall be documented.

🔍 **Rationale:** Documented procedures facilitate quick installation and support. Identification of installation schedule, resources, tools, and risks/mitigation upfront avoids delays. Documenting "lessons learned" and feeding this knowledge will avoid similar problems in future installations and will assist in root cause analysis activities.

☞ **TIPS:** Consider using elements of the design project plan (4.4.2.C.1) in developing this plan.

# HARDWARE/SERVICES ADDERS

The adders that are common to hardware and services are found in Sections 4.9 Process Control, and 4.10 Inspection and Testing. Also, there is a Note in Section 4.4.8 Design Validation that applies to hardware and service requirements. We identify these adders and the note in the next few pages, give you a rationale as to why they are a value-add to a quality system, and provide tips on implementing them.

## 4.4.8 Design Validation

🖫 **HV-NOTE J:*** See 4.10 for related requirements.

☞ **TIPS:** This note points out the parallel between design validation and inspection and testing of products.

## 4.9 Process Control

| 4.9.HV.1 Operational changes | 4.9.HV.2 Operator qualifications | 4.9.HV.3 Employee skills list |

---

*Notes are suggestions only not requirements.

**Adder: 4.9.HV.1: Operational Changes:** Each time a significant change is made in the established operation (for example, a new operator, new machine, or new technique), a critical examination shall be made of the first unit(s)/service(s) processed after the change.

**Rationale:** When significant changes are made, it is necessary to assure that the product is not affected. The process of thorough testing and examination will reveal if the change is free of defects and is free from negatively impacting the product.

**TIPS:** Records should be kept of when the change was made, the reason for the change, the impact, and the name of the individual who implemented the change.

Changes must be truly significant. For example, if a process is automated an operator change may not impact the quality of the output and a "first article" inspection may not apply. For installation service, a new installer, tool, or method must result in the examination of the initial outputs.

**Adder: 4.9.HV.2: Operator Qualifications:** The supplier shall establish operator qualification and requalification requirements for all applicable processes. These requirements, as a minimum, shall address employee experience, training, and demonstrated skills. The supplier shall communicate this information to all affected employees.

**Rationale:** Untrained, inexperienced, and underqualified employees can negatively impact quality and productivity. The qualification and experience required to perform each job at the supplier site needs to be documented. This will assist the operators in obtaining state-of-the-art training and remaining current in their knowledge for the job that they are performing. Any knowledge can become obsolete due to changing technology; therefore requalification is required to remain current.

**TIPS:** Consider education, experience, training, and demonstrated skills when identifying operator qualification. For installation service, this adder requires installer qualification and requalification criteria be established and communicated to employees.

**Adder: 4.9.HV.3: Employee Skills List:** The supplier shall maintain records of employees and their skills and qualifications (for example, employee skills bank or training records) to aid in determination of work assignments.

**Rationale:** In order to ensure that qualified individuals are assigned for the work, it is necessary to keep records of employee skills and experience. The records can be used to select the employee whose experience best suits the job to be performed. The skills list may also be used to determine training needs for individuals.

☞ **TIPS:** This adder requires that records be maintained, such as training and qualifications, and that they be used for work assignments.

## 4.10 Inspection and Testing

> **4.10.1.HV.1**
> Inspection and test documentation
>
> **4.10.5.HV.1**
> Inspection and test records

**Adder: 4.10.1.HV.1: Inspection and Test Documentation:** Each inspection or testing activity shall have detailed documentation. Details should include the following: (a) parameters to be checked with acceptable tolerances; (b) the use of statistical techniques, control charts, etc; (c) sampling plan, including frequency, sample size, and acceptance criteria; (d) handling of nonconformances; (e) data to be recorded; (f) defect classification scheme; (g) method for designing an inspection item or lot; and (h) electrical, functional, and feature testing.

**Rationale:** A well-defined test methodology which requires thorough test and inspection records will aid the testing activity. When each defect found in the test cycle is classified based on it's impact on the business, the number of defects can be used to determine the release of the product.

☞ **TIPS:** The defect data, when analyzed to determine the root cause, can be used to ensure that similar defects are prevented in the future.

The requirements in this adder are *shoulds*. You may use other procedures which can be shown to be equivalent to 4.10.1.HV.1.

**Adder: 4.10.5.HV.1: Inspection and Test Records:** Inspection and Test records shall include: (a) product identification; (b) quantity of product inspected; (c) documented inspection procedure(s) followed; (d) person performing the test and inspection; (e) date of inspection and/or test; and (f) number, type, and severity of defects found.

**Rationale:** Recording this information provides the supplier with the capability of tracing product from the field and/or repair back to the tests performed. This capability is a valuable tool for improvement of the inspection and testing process and the quality of the product.

☞ **TIPS:** This information is useful for both production and installation of the hardware. It's valuable to develop a culture of communication, feedback, and improvement linking product support, repair, and testing.

## HARDWARE ADDERS

The adders that are related to hardware only are found in Sections 4.4 Design Control; 4.8 Product Identification and Traceability; 4.9 Process Control; 4.10 Inspection and Testing; 4.11 Control of Inspection, Measuring and Test Equipment; 4.15 Handling, Storage, Packaging, Preservation, and Delivery; 4.19 Servicing; and 4.21 Quality Improvement and Customer Satisfaction. We identify these adders in the next few pages, give you a rationale as to why they are a value-add to a quality system, and tips on implementing them.

### 4.4 Design Control

🖙 **Adder: 4.4.4.H.1: Content of Requirements:** The design requirements shall include, but are not limited to: (a) nominal values and tolerances; (b) maintainability needs; and (c) end item packaging requirements.

🔍 **Rationale:** When the above elements are considered during the design phase, it ensures that the product's performance criteria are considered early in the product lifecycle.

☞ **TIPS:** Major characteristics to be considered are reliability, maintainability, maintenance support, and availability. This requirement may be covered as part of 4.4.4.C.2.

🖙 **Adder: 4.4.8.H.1: Periodic Retesting:** The supplier shall establish and maintain a documented procedure(s) that ensures products are periodically retested to assess the product's ability to meet design requirements.

# Hardware, Software, and Service Requirements 59

🔍 **Rationale:** Periodic retesting ensures that the product continues to perform according to the design requirements, and any changes made to the original design have not caused the product to deviate from the original design intent.

👉 **TIPS:** This should be included in the test plan (see 4.4.2.C.2). Document a procedure for testing individual units to original requirements and system-level summary tests.

💾 **Adder: 4.4.8.H.2: Content of Testing:** The initial test and periodic retest shall be more extensive than the routine quality control tests. The initial test shall include those that are contained in the customer and/or supplier product specifications and/or contracts. The results of these tests shall be documented.

🔍 **Rationale:** The initial tests will reveal the usability of the product which will be a determining factor for the release of the product.

👉 **TIPS:** This should be included in the test plan (see 4.4.2.C.2). Document a procedure for testing individual units to original requirements and system-level summary tests.

Vigorous testing of the initial product will uncover a majority of the defects. All the defects observed during testing must be documented, resolved, and recorded as quality records.

💾 **From H-NOTE L:**\* Product specifications may include environmental, vibration, flammability, and operational stress-type testing.

💾 **From H-NOTE M:**\* These documented procedure(s) should include:

a) Functional testing;

b) Qualification testing;

c) Stress testing;

d) Approved parts listing; and/or

e) Critical parts listing.

💾 **Adder: 4.4.8.H.3: Frequency of Testing:** The supplier shall establish and document the frequency for test/periodic retest. When determining the test frequency, the supplier shall include the following: (a) product complexity and service criticality; (b) number of design, engineering

---

\*Notes are suggestions only not requirements.

and/or manufacturing changes made to the product and whether the change(s) affect form, fit, and/or functions; (c) changes to the manufacturing process; (d) manufacturing variations (for example, tooling wear); (e) material and/or component substitutions and failure rates; and (f) the field performance record of the product.

**Rationale:** For thorough testing of the product, the frequency of tests must be determined on valid criteria, such as product complexity, frequency of the changes, and the number of changes to the product.

**TIPS:** This should be included in the test plan (see 4.4.2.C.2). Document a procedure for testing individual units to original requirements and system level summary tests.

**Adder: 4.4.9.H.1: Tracking Changes:** The supplier shall have adequate documented procedure(s) in place to ensure that material or component substitutions or changes do not adversely affect product quality or performance.

**Rationale:** Experience has shown that substitutions often have an adverse effect on the product. Changes may occur due to changing user requirements, hardware upgrades, or replacement of components.

Any failure of hardware will also require a change. Change management attempts to minimize problems by promoting identification of several factors such as dependencies, resources, and schedules and processes. Developing and documenting change control and substitutions procedure helps to manage the effects of change.

**TIPS:** Use a change management tool to track changes and maintain records which can later be reviewed and analyzed. This requirement can be covered as part of the change management process, 4.4.9.C.1.

**Adder: 4.4.9.H.2: Component Changes:** The supplier shall have adequate documented procedure(s) in place to ensure that material or component substitutions or changes do not adversely affect product quality or performance.

**Rationale:** It is very common for hardware components to be replaced by less expensive, equivalent components. When this happens, critical test and/or analyses must be made to assure that there is no effect on functionality, quality, or reliability.

**TIPS:** Use a change management tool to track changes and maintain records which can later be reviewed and analyzed. This requirement can be covered as part of the change management process, 4.4.9.C.1.

## 4.8 Product Identification and Traceability

> **4.8.H.1**
> Traceability for recall

> **4.8.H.2**
> Traceability of design changes

**Adder: 4.8.H.1: Traceability for Recall:** Field Replaceable Units (FRU) shall be traceable throughout the product lifecycle in a way that helps suppliers and customers to identify products being recalled, needing to be replaced, or modified.

**Rationale:** In case of a failure of previously released units, identification of products will assist in recalling the remaining units in a timely manner with less impact. Also, it will enable the supplier to identify those units that can be changed or modified in the field.

**TIPS:** This requirement can be included under the change management process, 4.4.9.C.1.

Use bar codes, serial numbers, CLEI (Telcordia provided) codes. Product support keeps track of units in the field based on customer inputs.

**Adder: 4.8.H.2: Traceability of Design Changes:** The supplier shall establish and maintain documented procedure(s) which provide traceability of design changes to identifiable manufacturing dates, lots, or serial numbers.

**Rationale:** Documenting design changes with dates, serial numbers, and lots will assist in recalling units in a timely manner with less impact. Also, it will enable the supplier to identify those units that can be changed or modified in the field.

This information will facilitate easy withdrawal or identification of field units when a defect is detected due to design changes.

**TIPS:** This requirement can be included under the change management process, 4.4.9.C.1. Also, link this requirement to 4.4.9.H.1, Tracking of Changes.

Use the engineering change order procedure. After approval there is a cutover date.

Manufacturing maintains serial numbers and tracks them for changes.

## 4.9. Process Control

> **4.9.H.1**
> Inspection and testing

**Adder: 4.9.H.1: Inspection and Testing:** Inspection and testing results shall be recorded and analyzed for the purpose of identifying problem areas.

**Rationale:** This is a corrective as well as preventive activity where the test results are documented and evaluated to identify the problem areas. The purpose is to detect the problems, analyze and correct them, and prevent similar problems from occurring in the future.

**TIPS:** Link this requirement to 4.10.5.HV.1 Inspection and Test Records.

## 4.10 Inspection and Testing

> **4.10.4.H.1**
> Testing of repair and return products
>
> **4.10.4.H.2**
> Packaging and labeling audit

**Adder: 4.10.4.H.1: Testing of Repair and Return Products:** Repair and return products shall be subjected to the same or equivalent documented final acceptance test procedure(s) as newly manufactured products.

**Rationale:** Repaired and returned products are expected to be the equivalent of new products. Therefore, they must go through the same rigorous testing procedure as new products. Failure of a unit the second time will have a bad effect on customer confidence.

**TIPS:** Ensure that there are no deviations from the documented testing procedures when it comes to returned or repaired products. This will help prevent return of repaired units a second time.

**Adder: 4.10.4.H.2: Packaging and Labeling Audit:** The supplier shall include a packaging and labeling audit in the quality plan or documented procedure(s). This may include, for example, marking, labeling, kitting, documentation, customer-specific markings, and correct quantities.

**Rationale:** The documented audit procedure will ensure effectiveness of the process related to packaging and labeling. Adequate packaging will prevent damage caused in shipping or handling. Labeling ensures correct identification of the product for installation, warranty, maintenance, recall, or repair.

**TIPS: H-NOTE R:*** This audit is normally done on products ready to ship.

## 4.11 Control of Inspection, Measuring, and Test Equipment

> **4.11.2.H.1**
> **Identified equipment**

**Adder: 4.11.2.H.1: Identified Equipment:** Inspection, measuring, and test equipment that is either inactive or unsuitable for use shall be identified and not used for production. All inspection, measuring, and test equipment that does not require calibration shall be identified.

**Rationale:** This process will prevent use of measuring and equipment that is not adequate for test or production and will also identify the equipment that does not require calibration. The purpose is to avoid the use of equipment that may give inaccurate or wrong results during production.

**TIPS:** Periodic audits of laboratory equipment is a means of assuring compliance to this requirement.

## 4.15 Handling, Storage, Packaging, Preservation, and Delivery

> **4.15.3.H.1**
> **Deterioration**

**Adder: 4.15.3.H.1: Deterioration:** Where the possibility of deterioration exists, materials in storage shall be controlled (that is, date stamped/coded) and materials with expired dates shall not be used.

**Rationale:** This requirement is aimed at identification of materials that if outdated may cause products to malfunction.

**TIPS:** Identifying such materials by date stamp or code will make it easy to track outdated materials, thereby preventing their use.

## 4.19 Servicing

> **4.19.H.1**
> **Supplier's recall process**

**Adder: 4.19.H.1: Supplier's Recall Process:** The supplier shall establish and maintain a documented procedure(s) for identifying and recalling products that are unfit to remain in service.

**Rationale:** Identification of resources to address field failures and support repair or exchange of product at customer sites will reduce the risk to the customer.

**TIPS:** Coordinate with or link this process to 4.8.H.1 Traceability for Recall and 4.8.HS.2 Product Identification.

## 4.21 Quality Improvement and Customer Satisfaction

> **4.21.3.H.1**
> **Field performance data**

**Adder: 4.21.3.H.1: Field Performance Data:** The quality system shall include the collection and analysis of field performance data which can be used to help identify the cause and frequency of equipment failure. In addition, no trouble found (NTF) data shall also be maintained. This information shall be provided to the appropriate organizations to foster continuous improvement. The quality system shall include a documented procedure(s) to provide the customer with feedback on their complaints in a timely manner.

*Rationale:* The data analysis of failures found in the field and identification of root cause of the failures will allow the implementation of preventive and corrective actions and should result in changes to processes to prevent similar failures in the future. Timely response to the customer complaints and inquiry will result in improved customer satisfaction. No trouble found (NTF) units are of special concern because they cause a loss of confidence in the diagnostic systems.

*TIPS:* Set up a database to gather field failure data and to link into a similar database containing repair and NTF data. Mechanisms should be set up to track NTF units returned to the field in order to prevent repetitive returns on the same board. If repair is done by a third party, the supplier or the customer should require repair and NTF data should be made available to the design organizations.

# SOFTWARE ADDERS

The software adders are found in sections 4.2 Quality System; 4.4 Design Control; 4.5 Document and Data Control; 4.9 Process Control; 4.10 Inspection and Testing; 4.15 Handling, Storage, Packaging, Preservation, and Delivery; and 4.19 Servicing. We identify these adders in the next few pages, give you a rationale as to why they are a value-add to a quality system and provide tips on implementing them.

## 4.2 Quality System

> **4.2.2.S.1**
> **Support software and tools management**

*Adder: 4.2.2.S.1: Support Software and Tools Management:* The supplier shall ensure that internally developed support software and tools used in the product lifecycle are subject to the appropriate quality method(s). Tools to be considered include design and development tools, testing tools, configuration management tools, and documentation tools.

**Rationale:** If the internally developed tools do not go through a rigorous validation, verification, and calibration process they may adversely effect the quality of the software.

**TIPS:** Use the same development and test procedures for design software and development tools that you use for development of production software. Place scripts for testing and problem tracking under configuration control.

## 4.4. Design Control

**Adder: 4.4.2.S.1: Estimation:** The supplier shall establish and maintain a method for estimating and tracking project factors during project planning, execution, and change management. Project factors to be considered should include product size, complexity, effort, staffing, schedules, cost, quality, reliability, and productivity.

**Rationale:** In project management, sound estimation techniques are crucial to the timely completion of the project. Poor estimation can result in cost and schedule overruns. Constant changes to the original product requirements will impact the schedule and cost. If the changes are not managed, the quality of the product will be affected.

**TIPS:** The estimation process should be ongoing during the design and development process. Use electronic tools to estimate resource needs throughout the project. Use inputs from other projects. Gather a body of knowledge.

**Adder: 4.4.2.S.2: Computer Resources:** The supplier shall establish and maintain methods for estimating and tracking critical computer resources for the target computer. Examples of these resources are utilization of memory, throughput, real-time performance, and I/O channels.

*Rationale:* The reliability, availability, and performance of the software depends on a thorough understanding of the resources needed by the deployed product. It's important during the design phase to continually estimate the utilization of memory, throughput, real-time performance and I/O channels in the target computer.

*TIPS:* The same tools used in 4.4.2.S.1 may be used to estimate the resources needed by the target computer. Use inputs from other projects. Gather a body of knowledge.

*Adder: 4.4.2.S.3: Integration Planning:* The supplier shall develop and document a plan to integrate the software components into the product. The plan shall include: (a) methods and documented procedure(s); (b) responsibilities; (c) schedule for integration; and (d) test requirements.

*Rationale:* Once all the modules (programs) are validated individually, the modules should be integrated and tested as a "whole system." When integrating the modules together, one module can have an adverse effect on another, data can be lost across interface or when combined together, or the modules may not produce the expected results. Integration testing will uncover the above mentioned "glitches" when the modules are combined together and tested to detect errors resulting from the integration.

*TIPS:* Use an integration test plan template which covers responsibilities, environmental requirements, and other elements of planning requirements.

*Adder: 4.4.2.S.4: Migration Planning:* When a system or software product is planned to be migrated from an old to a new environment, the supplier shall develop and document: (a) requirements analysis and definition of migration; (b) development of migration tools; (c) conversion of product and data; (d) migration execution; (e) migration verification; and (f) support for the old environment in the future.

*Rationale:* Migration from old to new environment must be documented to ensure that no functions are lost or degraded from the existing system when replaced with a new one. Migration planning should ensure that degradation of the reliability and performance of the software system does not take place.

*TIPS:* The migration plan should include components such as requirements analysis, tools which will be used in the migration activity, migrated data verification, and maintenance and support of the old environment during the migration process.

**Adder: 4.4.4.S.1: Identification of Software Requirements:** The supplier shall determine, analyze, and document the software component requirements of the system.

**Rationale:** A thorough requirements document affects the contents and functionality of a product. The requirements should be verified upfront to ensure that there is a common understanding of what is needed and how it will be accomplished. This understanding is necessary to both the supplier and the customer.

**TIPS:** Use requirements document reviews to uncover errors such as unclear requirements, missing functionality, or usability issues.

This requirement may be covered as part of 4.4.4.C.2 Design Requirements.

**Adder: 4.4.4.S.2: Requirements Allocation:** The supplier shall document the allocation of the product requirements to the product architecture.

**Rationale:** The management of the requirements during the design and architecture definition guarantees that every requirement is addressed in the development.

**TIPS:** Requirements allocation will ensure that the requirements are not "lost" as the project progresses from one phase to the next. Documenting the allocation will also help to ensure that appropriate test cases are written for the requirements, therefore giving full test coverage.

This requirement may be covered as part of 4.4.4.C.2 Design Requirements.

**Adder: 4.4.5.S.1: Design Output:** The required output from the design activity shall be defined and documented in accordance with the chosen method. Design outputs may include, but are not limited to: (a) architectural design, (b) detailed design, (c) source code, and (d) user documentation.

**Rationale:** The deliverables from the design—the design outputs—will assist in the maintenance of the software product during its life. These outputs will also facilitate testing and future changes.

**TIPS:** Use separate documents to cover (a) architectural design, (b) detailed design, (c) source code, (d) user documentation, etc. Outputs refer back to requirements in the parent document. One organization uses a template from their requirements planning document.

## 4.5 Document and Data Control

> **4.5.1.S.1**
> Control of customer-supplied documents and data

**Adder: 4.5.1.S.1: Control of Customer-Supplied Documents and Data:** The supplier shall establish and maintain documented procedure(s) to control all customer-supplied documents and data (for example, network architecture, topology, capacity, and database) if these documents and data influence the design, verification, validation, inspection and testing, or servicing of the product.

**Rationale:** The customer may need to provide documents and data to assist in the development of the product. The documents and data may contain "proprietary" information and failure to institute proper controls of identification can result in legal problems and loss of customer confidence.

**TIPS:** The same procedures used for control of internal documents and data may be used for this requirement.

## 4.9 Process Control

> **4.9.S.1** Replication     **4.9.S.2** Release management

**Adder: 4.9.S.1: Replication:** The supplier shall establish and maintain a documented procedure(s) for replication which should include the following: (a) identification of the master copy and copies to be delivered; (b) the number of copies to be delivered; (c) type of media and associated labeling; (d) identification and packaging of required documentation such as user manuals; and (e) controlling the replication environment to ensure repeatability.

**Rationale:** The customer may have several sites and may require the supplier to create and deliver more than one copy of the product. This adder ensures that there are no defects introduced during labeling, packaging, and duplication. Identification of the master copy will prevent the risk of duplicating from an obsolete copy.

☞ **TIPS:** If replication is done at another facility, the design organization should retain a "gold" copy and send a "silver" copy to the replication center. The replication center can then send a sample back to the design organization for validation testing against the gold copy.

**Adder: 4.9.S.2: Release Management:** The supplier shall establish and maintain documented procedure(s) to control the release and delivery of software products and documentation. These documented procedure(s) should include methods to provide for the following: (a) release planning information delivered to the customer sufficiently in advance of the release; (b) product introduction and release schedules to the customer; (c) detailed description of product features delivered, and changes incorporated in new software products of releases; and (d) advising the customer of current or planned changes.

**Rationale:** To control the delivery of a product, development of a release plan is necessary. The intent of this document is to describe the contents of the release and the flow of the engineering release process. Entrance and exit criteria for the various stages of the release process are also described.

At the same time, the customer needs time to prepare to receive the new product, allocate resources to support it, and schedule training for the staff. Receiving a detailed content list of the product features will assist the customer in conducting acceptance testing and in ensuring that the product has all the functions that were stated in the requirements document.

☞ **TIPS:** The project plan described in 4.4.2.C.1 can be used as a guide in planning the release and delivery of software products and documentation.

Some inputs to be considered at time of release:

- Product change notices;
- Customer information reports;
- Beta tests with customers;
- Customer documentation tests;
- Release notes with exceptions;
- Products that go to customer labs first.

# Hardware, Software, and Service Requirements 71

## 4.10 Inspection and Testing

> **4.10.1.S.1**
> Test documentation

**Adder: 4.10.1.S.1: Test Documentation:** Software tests shall be conducted according to documented procedure(s) and the test plan. Documentation of testing shall include: (a) test results, (b) analysis of test results, (c) conformance to expected results, and (d) problem reporting for nonconforming items.

**Rationale:** Testing is a complex engineering activity and must be planned in order to have effective results. The supplier must identify how the product will be tested, the types of test to be conducted, document actual results, and the expected results of the tests. If the expected results were not identified in each test case, how would one know whether the test has passed or failed? Any failed tests results must be documented so they can be fixed and retested prior to the release of the product.

**TIPS:** Use a rigorous system test procedure in which test cases are developed from the requirements by an organization independent of the design organization. System testers should be challenged to "break the system."

Document system test and validation test plans. Reports should have test results, conformance to expected results, and a list of nonconforming items.

## 4.15 Handling, Storage, Packaging, Preservation, and Delivery

> **4.15.2.S.1**
> Software virus protection
>
> **4.15.6.S.1**
> Patch documentation

**Adder: 4.15.2.S.1: Software Virus Protection:** The supplier shall establish and maintain methods for software virus prevention, detection, and removal from the deliverable product.

**Rationale:** In today's environment, we have seen the negative results of virus attacks. A virus can bring down entire businesses and result in the loss of millions of dollars. In order to prevent the customer from encountering the ill effects of any type of virus, the supplier must establish processes to protect against unwanted penetration of the virus.

**TIPS:** Develop a procedure on how to create disks and CDs and for ensuring that no virus exists. Test each disk before sending them.

**Adder: 4.15.6.S.1: Patch Documentation:** The supplier shall establish and maintain methods to ensure that all documentation required to describe, test, install, and apply a patch has been verified and delivered with the patch.

**Rationale:** Patches are necessary to address emergency fixes that are supplied to the customers to address specific defects. The supplier must establish a process to test these emergency fixes to make certain they work and do not create adverse effect on the existing system. A customer's ability to install the patch smoothly without encountering difficulties depends on the process used by the supplier to test the patch prior to the delivery. Verification of patch-related documentation will help ease customer use and assure accuracy.

**TIPS:** See 4.19.S.1 for related information.

## 4.19 Servicing

*4.19.S.1 Patching procedures*

*4.19.S.2 Problem resolution*

**Adder: 4.19.S.1: Patching Procedure(s):** The supplier shall establish and maintain documented procedure(s) that guide the decision to solve a problem by patching in the following manner: (a) The documented procedure(s) shall address patch development procedures, propagation (forward and backward), and resolution; (b) This documented procedure(s) shall be consistent with purchaser needs or contractual requirements for maintenance support; (c) For each patch, the supplier shall provide the customer with a statement of impact regarding that patch on the customer's operation.

**Rationale:** Often there is a dilemma regarding whether to send a patch to a customer for resolving defects or to send a product update.

## Hardware, Software, and Service Requirements

Depending on how many changes/fixes are addressed in a patch and the impact of the changes/fixes to the customer the supplier should establish a procedure to identify when a patch will be sent versus an update.

**TIPS:** See 4.15.6.S.1 for related information covering the documentation of patches.

**Adder: 4.19.S.2: Problem Resolution:** The supplier shall establish and maintain documented procedure(s) to initiate corrective action once a reported trouble is diagnosed as a problem. The documented procedure(s) should provide guidelines for distinguishing among potential solutions such as: (a) patching; (b) immediate source code corrections; (c) deferring solutions to a planned release; and (d) providing documented "work-around" operational procedure(s) and resolution within a designated timeframe based on the severity of the issue.

**Rationale:** A great deal of time is wasted in analyzing and fixing similar problems that continue to plague every release of the software. By having a structured process for conducting root cause analysis and implementing corrective and preventive actions, the supplier may succeed in eliminating the "same type" of defects from reoccurring in the future.

**TIPS:** This requirement may be satisfied in conjunction with 4.19.S.1. The method of addressing defects by either using patch, immediate source code corrections, deferring solutions for a future planned release, or providing a "work-around" is common to satisfying the two requirements.

## SERVICE ADDERS

The adders that are related to service only are found in Sections 4.4 Design Control, 4.9 Process Control, and 4.21 Quality Improvements and Customer Satisfaction. We identify these adders in the next few pages, give you a rationale as to why they are a value-add to a quality system, and provide tips on implementing them.

## 4.4 Design Control

**V-NOTE D:*** **Design of Services:** All design control requirements apply to suppliers that are responsible for the design of delivered services.

**TIPS:** This note was added to insure that services design suppliers are aware that all the design elements and design adders are applicable to designed services.

**Adder: 4.4.5.V.1: Services Design Output:** The required output from the services design shall contain a complete and precise statement of the service to be provided. Design outputs shall include, but are not limited to: (a) service delivery procedures; (b) resource and skill requirements; (c) reliance on subcontractors; (d) service characteristics subject to customer evaluation; and (e) standards of acceptability for each service characteristic.

**Rationale:** In order to expedite the service delivery with minimal impact on the customer, the supplier must have documented agreements with the customers that outline the type and extent of the service to be provided. In addition, the supplier must have trained staff with thorough understanding of the service delivery procedures. The acceptance criteria for the service and the standards of acceptability must be identified and documented as a part of the service contract to avoid any misunderstanding at a later time between the supplier and the customer.

**TIPS:** The following list is relative to installation services only:

a. Requires that installation manuals be developed;

b. Requires that a listing of tools and test sets and installer skills be developed;

c. Requires a listing of noninstaller operations be developed;

d. Requires a listing of items to be evaluated by the customer;

e. Includes a listing of requirements.

**Adder: 4.4.9.V.1: Tool Changes:** The supplier shall have documented procedure(s) in place to ensure that substitutions or changes to tools used in performing the service do not adversely affect the quality of the service.

**Rationale:** Changes to tools used in performing the service can adversely affect the quality of the service.

---

*Notes are suggestions only not requirements.

☞ **TIPS:** It is important to document a structured process describing how tool changes will be reviewed for their effect on the service. This should include operational testing of the tools before actual use.

## 4.9 Process Control

▦ **Adder: 4.9.V.1: Software Used in Service Delivery:** Suppliers shall document and implement processes for the maintenance and control of software used in service delivery to ensure continued process capability and integrity.

🔍 **Rationale:** Delivery of some services requires the use of software tools. Examples are tools used to download the software being installed and tools to do diagnostics during the installation. These tools can have defects that will affect their ability to deliver the service. Identifying and resolving defects will assist in minimizing service interrupts in the field due to malfunctioning software.

☞ **TIPS:** The organizations responsible for service should have a documented procedure for verifying and validating the software used in service delivery. The methodology for verifying and validating the software is similar to the methodology for verifying and validating the delivered software.

▦ **Adder: 4.9.V.2: Service Delivery Plan:** Suppliers that are responsible for the delivery or implementation of a service, and are not responsible for the design of that service, shall comply with the project plan requirements of 4.4.2.C.1.

🔍 **Rationale:** Suppliers not responsible for the design of a service are not required to satisfy element 4.4. This necessitated the creation of adder 4.9.V.2 which requires use of a project plan as defined in 4.4.2.C.1. This will assist in controlling slippage of schedule and will identify issues before they impact the project adversely.

☞ **TIPS:** Since 4.4.2.C.1 contains *should* requirements, some of the items may not be required for service delivery. Be sure to document why they are not required.

4.4.2.C.1 C-NOTE E states that "the project plan and related plans may be an independent document, a part of another document, or composed of several documents." This note provides the supplier the latitude to use the same document(s) for several different projects.

## 4.21 Quality Improvement and Customer Satisfaction

**Adder: 4.21.3.V.1: Service Performance Data:** The quality system shall include the collection and analysis of service performance data, which can be used to identify the cause and frequency of service failures. This information shall be provided to the appropriate organizations to foster continuous improvement of the service.

**Rationale:** The quality system is a closed-loop process where measurements must be taken to determine the quality of products, processes, and services. Service failure data should be used for implementing corrective and preventive actions. The performance data gives you information to determine where improvement can be made in the process.

**TIPS:** This adder requires that there be a system in place to collect services related problems, and that the data be analyzed and fed to the appropriate organizations for use in their quality improvements.

# END NOTES

1. ANSI/ISO/ASQC Q9001–1994, *Quality Systems—Model for Quality Assurance in Design/Development, Production, Installation, and Servicing*, Second Edition (Milwaukee, WI: American Society for Quality Control, 1994).
2. *TL 9000 Quality System Requirements*, Book One, Release 2.5 (The QuEST Forum, 1999).
3. GR-1202-CORE, *Generic Requirements for Customer Sensitive Quality Infrastructure* (Morristown, NJ: Telcordia Technoligies, Issue 1, October 1995).
4. SR-3535, *Bellcore CSQPS$^M$ Program* (Morristown, NJ: Telcordia Technologies, Issue 1, October 1995).

# 4
# TL 9000 Metrics

## THE IMPORTANCE OF METRICS

One of the fundamental drivers for creating the QuEST Forum and TL 9000 was the telecommunication industry's inability to measure the impact of improvements on the cost of quality. Both service providers and suppliers were helpless in this regard. Another motive was to measure in real terms the effect of the requirements and the metrics applied to the telecom industry. The Metrics Handbook is intended to help guide progress and evaluate results of the TL 9000 quality system implementation.

Another major driver behind measurement was the following: *Unless we know where we are, there is no way we can define goals and objectives for future improvement*. With the solid push for continuous improvement, it was imperative that TL 9000 adopt a metrics standard.

All business excellence criteria place extensive emphasis on measurements. The Malcolm Baldrige National Quality Award (MBNQA) Criteria for Performance Excellence[1] states: "... *measures* ... represent factors that lead to improved customer, operational, and financial performance. A comprehensive set of *measures* ... represents a clear basis for aligning activities with the company's goals," and " ... (a balanced *set of measures*) ... offers an effective means to communicate ... priorities, to monitor actual performance, and to marshal support for improving results.".

The European Foundation for Quality Management (EFQM) model[2] states the importance of measurements as "*Management by Facts* ..." ... and "... EFQM model is a practical tool to help organizations do this by *measuring* where they are on path to Excellence."

**78** Chapter Four

There can be no evaluation without metrics. Take the example of any sport activity, which are meaningless without some kind of measurement. For some reason, there is a negative connotation attached to the term *metrics*, as some people think that the sole purpose of having metrics is to compare their performance against their peers or competitors. People fail to realize the broader important perspective behind metrics collection.

In the telecommunication environment, in particular, metrics are beneficial for the following reasons:

- to determine whether stated goals and expected benefits of the TL 9000 quality system requirements are being achieved;

- to protect integrity and enhancement of value of telecommunications products and services;

- to achieve continuous improvement of products and subscriber services;

- to enhance customer-supplier relationships;

- to improve subcontractors' performance;

- to see overall cost reduction and increased competitiveness of suppliers and service providers;

- to show value of quality and reliability successes of TL 9000 participants, and encourage wider TL 9000 implementation and acceptance;

- to help direct quality system management activity to areas where the most beneficial results will be realized;

- to show all industry stakeholders (customers, employees, stockholders, suppliers, the public, etc.) how their interests are being addressed; and

- to determine performance/quality relative to others in the industry.

## INTRODUCTION

The uniqueness of the TL 9000 quality standard lies in its requirement for metrics submission. No other known quality standard in any industry has a similar requirement. The main purpose for metrics submission is to make available a few vital measurements by which suppliers can judge their own performance in relation to the industry norm. A long-term purpose is to create an index for the telecom industry by which the industry as a whole can measure its progress.

As anticipated, there was initially some resistance towards this move for measurements. But it soon wilted because everyone realized that the improvements cannot be measured without the metrics and metrics can only help everyone in the industry.

## METRICS WORK GROUP

The next question was how to go about defining which metrics to include in the handbook. A team of quality experts with background in measurements was chartered to define the metrics that would be part of the handbook. The team set out to perform their new assignment with enthusiasm. They soon realized the enormity of the task at hand and decided to hold monthly meetings for the creation of the metrics handbook.

Initially the discussions were lengthy and sometimes heated. They lingered on what type of metrics should be considered. The choice was between product metrics, process metrics, or both. The service providers wanted both product and process metrics to be included. On the other hand, the suppliers justifiably argued that if too many metrics were included, the cost of collecting, reporting, and monitoring them would be very high and would eventually derail the overall goal of cost reduction.

After the debates lasted for a few sessions, the metrics work group approached the handbook creation process very professionally. To avoid further arguments, the team first decided on the following rules and guidelines for the development of TL 9000 metrics.

---

**Metrics Ground Rules**

1. The metrics can fall in the following categories:

   (a) Hardware; (b) Software; (c) Services; or (d) Common (to Hardware, Software, and Services).

2. Each measure has a defined audience, a defined purpose, is consistent, objective, easily understood, and useful for evaluation or decision making.

3. Each measure provides data to support timely analysis to determine trends, projections, or cause and effect.

4. Use common basis (normalized) for similar products/services—making inevitable comparisons as fair as possible.

5. Select a small number per category.

6. Use or modify existing metrics where possible—e.g., service provider report cards, RQMS (GR 929-CORE),[3] IPQM (GR-1315-CORE),[4] Supplier Data (GR-1323-CORE).[5]

7. Defined as potential part of a possible future industry index (J.D. Power-like).

8. Address different product types (e.g., cable, switch, outside plant hardware, switch systems/software, cross connect systems/software, etc.) and defined phase(s) of their lifecycles.

Based on the above guidelines, a list of criteria was developed which would make the metrics package acceptable. The criteria were:

1. Well-defined counting rules for each metric;

2. Product applicability rules based on a well-defined, unambiguous, product category table;[*]

3. Clear, auditable, collection/reporting processes; and

4. Unambiguous answers to any question that may arise with respect to:
   - logistics associated with suppliers collecting data from each subset of their customers;
   - logistics surrounding supplier data aggregation;
   - logistics regarding a geographically dispersed supplier where not all metrics for a given product can be reported by a given geographic organization;
   - logistics regarding the relation between all suppliers in the supply chains and their role in metrics collection/reporting;
   - logistics and industry ramifications regarding published benchmark data.

The above criteria were effectively used in metric definition.

---

[*]The Product Category Table can be viewed on the QuEST Forum website: www.questforum.org.

# METRICS SELECTION USING THE GOAL-QUESTION-METRIC MODEL

The metrics work group decided to adopt the goal-question-metric (GQM) model to determine which metrics to select. The GQM paradigm explains a measurement and evaluation method,[6] which entails asking questions to help identify goals. For each goal, ask questions that will help one understand whether the goal has been accomplished. Many questions have answers that can be measured.

The GQM paradigm is a useful tool in identifying good, appropriate metrics. It entails three aspects:

1. What are the goals of the metrics?

2. What questions would customers and suppliers like answered by these metrics?

3. What are the metrics that would answer these questions?

The GQM exercise led to the list of the following goals.

---

**Goals of Metrics**

1. Reduce the cost of ownership
   1A Reduce cost of ownership by the Purchaser
   1B Reduce cost of ownership by the Supplier

2. Increase customer satisfaction—confidence

3. Increase predictability (on-time delivery, delivered functionality, cost, final product quality)

4. Improve current performance (product & process):
   4A Current performance—Product related
   4B Current performance—Process related

5. Improve lifecycle performance:
   5A Improve lifecycle performance—Process related
   5B Improve lifecycle performance—Product related

6. Reduce cost impact of software faults (supplier & customer)—covered by #1A & #1B

7. Improve quality—covered by #5B

8. Reduce cycle times for software delivery

The next step involved creating a set of questions for each goal. The following list of questions is a sample to demonstrate the approach taken by the group. Sample questions are provided for goals 1A, 1B, 2, 5B, and 8.

---

**Sample of Questions**

1A. Reduce the cost of ownership—by the purchaser

- How much does it cost the Purchaser?
  - To acquire
  - To install/implement
  - To operate (day to day—normal conditions)
  - To maintain (corrective—cost of failure)
  - To grow (adaptive)
  - To retire

1B. Reduce the cost of ownership—by the supplier

- How much does it cost the Supplier?
  - To acquire/develop/enhance (concept—delivery)
  - To manufacture/deploy/install
  - To support (day to day—normal conditions)

2. Increase customer satisfaction—confidence

- How satisfied are our customers?
- What factors do our customers find important?
  - On-time delivery
  - Product performance
  - Product support
- Product introduction
- Technology leadership (perception)
  - How satisfied and confident are our customers with:

- Our ability to meet commitments
- The quality of our products & services
- Our responsiveness to customer requests
- The fix response time (responsiveness)

5B. Improve released product quality (service provider)
- How many field failures were found by severity?
- How many field outages occurred (events & duration)?
- Did the delivered products meet the requirements?
  - Feature functionality
  - Interoperability
- What is the "out-of-the-box" quality (installability, initial operation, documentation, completeness of delivery—may be a system issue)?
- What areas of the software have the highest concentration of defects?
- What is the quality of the software fixes?

8. Reduce cycle times for SW delivery
- How much time is spent in each development phase and activity?
- What are the bottlenecks in the supplier's process?
- What is the overall cycle time release to release?
- How much functionality is delivered within a given cycle window (production rate)?
- How effectively are resources being planned and used?

In view of the enormous task of defining metrics, the group decided to develop the metrics into two parts. Part I would include field metrics or the metrics collected after a product is delivered and installed in the field. Part II would include process metrics, which would exhibit the development process efficiency and effectiveness.

## PHASE 1 METRICS

The list of metrics from Phase I is divided into four groups: a) Common: applicable to Hardware, Software and Service categories; b) Hardware: applicable to Hardware products only; c) Software: applicable to Software products only; and d) Service: applicable to Service category only. The Phase 1 metrics are listed in the following exhibit. The detailed descriptions of metric counting rules, calculations, formulas, and example calculations are provided in the Metrics Handbook.[7]

---

**Phase 1 Metrics**

- Common
    1. Number of problem reports
    2. Problem report fix response time (2A) and overdue fix responsiveness (2B)
    3. On-time delivery
    4. System outage
- Hardware
    1. Circuit pack return rates
- Software
    1. Software update quality
    2. Release application aborts
    3. Corrective patch quality (3A) and feature patch quality (3B)
- Service
    1. Service quality

---

All suppliers are expected to establish a metrics system through which metrics collection, monitoring, and reporting takes place. The unique aspect of this system is that service providers also have a major role to play, in that for some metrics, service provider cooperation is a must. Unless they provide field data to the suppliers, the suppliers cannot generate certain metrics. The Metrics Model in Figure 4.1 depicts at a higher level the cooperation between suppliers, service providers, and the QuEST Forum.

Figure 4.1 shows the flow and usage of TL 9000 metrics as described in the handbook.[7] It explains the interaction between service providers, suppliers, and the Metrics Repository System (MRS). The three must work in tandem to make Metrics collection, reporting, and analysis a reality. In addition, contractually required metrics are included to provide the complete picture as to how metrics will be used to drive continuous improvement in the telecommunications industry.

> **Explanation of Figure 4.1**
>
> 1. Service provider must provide the supplier with data for some metrics.
>
> 2. The supplier calculates the metrics and provides them to the MRS at the University of Texas at Dallas (UTD).
>
> 3. UTD uses metrics tools (3A) to calculate statistics (3B) by product class.
>
> 4. UTD calculations are posted on the QuEST Forum web site.
>
> 5. Inputs are provided to the QuEST Forum about continuous improvement programs from external data (2), QuEST Forum web site (4), internal TL 9000 metrics (5A), and assessment report/report cards (5B).
>
> 6. Service provider may require other data from the supplier.

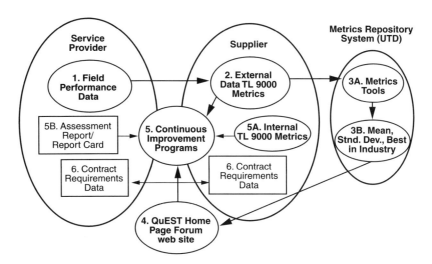

**Figure 4.1** QuEST TL 9000 Metrics Model.

## METRICS USAGE

There are three primary uses of the metrics. They are:

1. To report to Metrics Repository System (MRS);
2. For internal use in continuous improvement programs; and
3. In customer/supplier relationships.

The following principles for processing and usage of the metrics are meant to be consistent with the purpose of providing an environment where service providers and suppliers can work together to drive continuous improvement:

- Suppliers will provide TL 9000 metrics to UTD who will compile all the metric data and calculate an "industry mean," "standard deviation," and "Best in Industry" for each category. Results or reports produced by the UTD will not identify individual suppliers.

- Service providers may request their suppliers to provide the TL 9000 metrics specific to that service provider. This information exchange will take place strictly between the supplier and the service provider per contractual agreements. The Forum and Metric Administrators will not be involved in any way.

- Suppliers will not be ranked by UTD.

The general uses of the metrics currently envisioned are:

- "Industry Mean," "Standard Deviation," and "Best in Industry" statistics will be used by suppliers to improve products and services. The metrics provide data that is needed by suppliers to identify significant improvement areas. After improvement actions have been taken, the data will allow suppliers to determine the level of improvement and decide if additional significant improvement is needed.

- "Industry Mean," "Standard Deviation," and "Best in Industry" statistics along with supplier specific metrics may be used by service providers to evaluate and work with suppliers. By comparing a supplier's performance against the industry data, gaps in performance can be determined and the service provider and supplier can address closing the gaps.

- Suppliers and service providers will use the metrics to determine if products, intercompany processes, and services meet end user expectations.

- To improve the industry products and services based on Forum initiatives and goals.
- To harmonize the supplier assessment process.

After the process and metrics have matured:

- They will be used to establish a Telecommunication Quality Index to be published worldwide.
- The Forum may create an "award program" for best in class.

The text below provides an overview of the Part I metrics. These metrics are product related.

## COMMON METRICS

Common metrics are applicable to the Hardware, Software, and Service categories. There are four common metrics: (1) Number of Problem Reports; (2A) Problem report fix response time; (2B) Overdue fix responsiveness; (3) On-time delivery; and (4) System outage.

### Number of Problem Reports (NPR)

The number of problem reports (NPR) is computed as the total number of problem reports per normalized unit per month. This metric is adopted from RQMS[3] and applies to all products.

The purpose of this metric is to evaluate the number of problem reports or complaints during field operations in order to reduce their number along with associated costs and revenue losses. NPR provides a common means whereby customers and suppliers can evaluate the rate of problem reports, including engineering complaints (EC), that are reported on the supplier's product. This would induce joint customer-supplier efforts to reduce costs and revenue losses associated with problems.

Monitoring the trends of problem reports helps the customer understand whether the impact of product quality on the field is improving as measured by reported problems. It is a measure of the reported field "pain" experienced by the customer.[3]

Only customer reported problem reports are counted. Both customers and suppliers provide data for NPR measurement. Since the origin of this metric is in the RQMS, an alternative way of reporting is also described in the Handbook.

One report is to be compiled per supplier product category.[8] Typical target audiences for NPR data are customer maintenance, engineering, and

Problem Reports per System—Six Month Moving Average

|  | Mar-99 | Apr-99 | May-99 | Jun-99 | Jul-99 | Aug-99 | Sep-99 | Oct-99 | Nov-99 | Dec-99 | Jan-00 | Feb-00 | Mar-00 |
|---|---|---|---|---|---|---|---|---|---|---|---|---|---|
| Critical | 0.000 | 0.000 | 0.000 | 0.000 | 0.040 | 0.033 | 0.033 | 0.033 | 0.033 | 0.033 | 0.000 | 0.017 | 0.017 |
| Major | 0.000 | 0.000 | 0.033 | 0.025 | 0.060 | 0.100 | 0.117 | 0.117 | 0.133 | 0.133 | 0.100 | 0.067 | 0.067 |
| Minor | 0.200 | 0.250 | 0.333 | 0.375 | 0.440 | 0.450 | 0.550 | 0.650 | 0.700 | 0.650 | 0.550 | 0.500 | 0.417 |
| Systems in Service 100's | 10.00 | 12.50 | 15.00 | 13.75 | 11.60 | 10.33 | 11.71 | 11.00 | 10.33 | 9.60 | 9.27 | 9.17 | 9.38 |

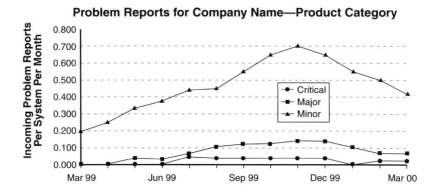

**Figure 4.2** Total problem reports by severity level.

technical support, supplier technical support, manufacturing management, and engineering management.

The metric to be reported to UTD must be in a tabular form as described in the Handbook. A graphical form depicting the trend of minor, major, and critical problem reports is illustrated in Figure 4.2.

## Fix Response Time (FRT), Overdue Problem Report, and Fix Responsiveness (OFR)

Fix response time (FRT) represents the responsiveness of suppliers to a subset of customer reported problems that require a change to the product. FRT measures the overall responsiveness of the supplier to all major and minor problems reported in hardware and software categories and all problems reported under service category. Overdue problem report fix responsiveness (OFR) measures the rate of closure of overdue major and minor hardware and software problems and all service problem reports. Both metrics are adopted from RQMS.[3]

The problem report fix metric for each product category includes the following:

- Percent Hardware/Software major problems fixed within 30 days;
- Percent Hardware/Software minor problems fixed within 180 days;
- Percent service problems fixed on time.

The intent is to get the suppliers to respond to problem reports in an efficient and effective manner, so as to facilitate prompt availability. Critical problem reports are excluded because they are worked exclusively until their resolution.

The fix response metric monitors the supplier's responsiveness to problem reports that for resolution may require changes to the product, hardware fixes, document changes, procedure changes, and/or software fixes (patches, S/W upgrades or code changes in future generics). The calculation measures the percentage of fixes due to be made available that meet the closure targets.

The overdue fix responsiveness calculates percentages from the number of problem reports (PRs) overdue, the number of penalty PRs (for non-service problems) and the number of overdue PRs closed each month for both major and minor severity reports.

The data for this metric is provided by the service providers. Figures 4.3 and 4.4 are examples of the overall fix response time and overdue fix responsiveness. Figure 4.3 includes data on critical problems.

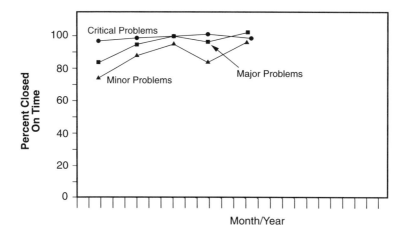

**Figure 4.3** Overall fix response time.

**Figure 4.4** Overdue fix responsiveness.

## On-Time Delivery (OTD)

The purpose for measuring OTD is to evaluate the supplier's on-time delivery performance in order to meet the customer's need for timely product and service delivery and to meet end customer expectations.

There are three metrics identifiers for OTD. They are: (1) OTIS (On-Time Installed System Delivery); (2) OTI (On-Time Items Delivery); and (3) OTS (On-Time Service Delivery). OTD is measured against customer requested date (CRD) for each of three distinct cases (systems, items, and services).

- OTIS: Installed System Order Fulfillment—OTD for installed systems is computed as follows:
  Percent (%) Systems accepted on CRD = # Systems accepted on CRD divided by # Systems for which CRD occurred during the reporting period.

- OTI: Orderable Line Item Order Fulfillment—OTD for orderable line items is computed as follows:
  Percent (%) Orderable Line Items delivered on CRD = # of Orderable Line Items delivered complete on CRD divided by # of Orderable Line Items for which CRD occurred during the reporting period.

- OTS: Service Order Fulfillment—OTD for service orders is computed as follows:
  Percent (%) Service Orders delivered on CRD = # of Service Orders delivered complete on CRD divided by # of Service Orders for which CRD occurred during the reporting period.

**Table 4.1** Example of orderable line item OTD.

| | PO | Line Item | Qty | CRD | Qty | Actual | OTD CRD | NOTE |
|---|---|---|---|---|---|---|---|---|
| | A | 1 | 5 | 3/10 | 5 | 3/10 | 1 | |
| | | 2 | 6 | 3/12 | 6 | 3/13 | 0 | |
| | | 3 | 4 | 3/17 | 4 | 3/18 | 0 | |
| | B | 1 | 8 | 3/20 | 8 | 3/22 | 0 | |
| | | | | | | 3/23 | | |
| | | 2 | 12 | 3/22 | 6 | 3/22 | 0 | 4 |
| | | | | | 6 | 3/25 | | |
| | | 3 | 2 | 3/29 | 2 | ? | 0 | 6 |
| | | 4 | 2 | 3/30 | 2 | 3/29 | 0 | 5 |
| | | | | | | 3/30 | | |
| | C | 1 | 7 | 2/15 | 7 | 3/15 | NA | 2 |
| | | 2 | 1 | 2/15 | 1 | 3/15 | NA | 2,3 |
| TOTAL | 3 | 9 | 47 | | 7 | | 1 | |
| March | | OTD | | | | | 14% | 1 |

Notes:
1. The CRD OTD performance for March was 14% or 1 (CRD met)/7 (CRDs due).
2. PO line items C1 and C2 CRDs were not counted in the total of 7 for March because they had February CRD dates.
3. PO line item C2 SCD was not counted because it had a February date.
4. Line item B2 was not on time for CRD because only 6 of the items were delivered to CRD.
5. Line item B4 was given no credit for partial on time delivery. In addition, it would not be given credit because of early delivery of 1 unit.
6. Line item B3 has "?" for the Actual delivery date since this item was not delivered by the end of March when this report was prepared.

**Table 4.2** Example of installed system OTD.

| | PO | CRD | Line Item | Qty | Actual | Acceptance | OTD CRD | Note |
|---|---|---|---|---|---|---|---|---|
| | A | 3/10 | 1 | 5 | 3/10 | | 1 | |
| | | | 2 | 6 | 3/10 | | | |
| | | | 3 | 4 | 3/10 | 3/10 | | |
| | B | 3/20 | 1 | 8 | 3/22 | | 0 | |
| | | | | | 3/23 | | | |
| | | | 2 | 12 | 3/22 | | | |
| | | | | | 3/25 | 3/25 | | |
| | C | 3/21 | 1 | 2 | 3/21 | | 0 | |
| | | | 2 | 2 | 3/21 | | | |
| | | | | | 3/22 | 3/22 | | |
| | D | 2/15 | 1 | 7 | 3/15 | | 0 | 2 |
| | | | 2 | 1 | 3/15 | 3/15 | | |
| TOTAL | 4 | 3 | 9 | 47 | | | 1 | |
| March | OTD | | | | | | 33% | 1 |

Notes:
1. The CRD installed OTD performance for March was 33% or 1 (CRD met) divided by 3 (CRDs due).
2. PO system D CRDs were not counted in the total of 3 for March because they had a Feb. CRD date.

## 92 Chapter Four

| | Date | 1996 | 1997 | Jan | Feb | Mar | Apr | May | Jun |
|---|---|---|---|---|---|---|---|---|---|
| TYPE | Met CRD | 15 | 22 | 3 | 2 | 1 | 2 | 2 | 3 |
| Systems | Total | 40 | 45 | 5 | 4 | 3 | 5 | 4 | 5 |
| | CRD | 38% | 49% | 60% | 50% | 33% | 40% | 50% | 60% |
| | LOR | 6 | 5 | 1 | 0 | 0 | 0 | 0 | 0 |
| Line Items | Met CRD | 18 | 22 | 2 | 3 | 1 | 3 | 4 | 4 |
| | Total | 45 | 50 | 5 | 6 | 7 | 7 | 8 | 7 |
| | CRD | 40% | 44% | 40% | 50% | 14% | 43% | 50% | 57% |
| | LOR | 8 | 6 | 0 | 1 | 0 | 0 | 0 | 0 |

LOR = Late Order Received

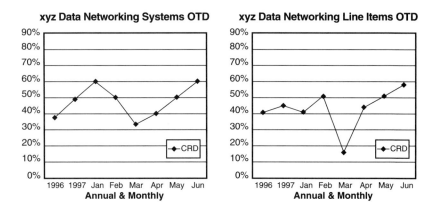

**Figure 4.5** Example charts and graphs for on-time delivery metrics.

Tables 4.1 and 4.2 illustrate computation of OTD metrics from a series of deliveries of orderable line items/installations of systems for purchase orders (PO). Reports of OTD data may be presented in a table and chart as shown in Figure 4.5. The acronyms used in the examples are customer requested date (CRD), purchase order (PO), and quantity (Qty).

## System Outage

The purpose of measuring system outage is to evaluate the downtime performance and outage frequency during field operation in order to reduce both the frequency and duration of outages and their associated cost and revenue impacts.

System outage applies to Hardware and Software products and is a measure of complete loss of functionality of all or part of a system. System outage downtime metric varies by product. The System outage frequency metric is outages per normalization unit per year.

The system outage metric is computed for the following aspects:

- Annualized outage frequency
- Annualized downtime
- Annualized supplier attributable outage frequency
- Annualized supplier attributable downtime

An example of system outage is presented in Table 4.3. It contains an example of overall TL 9000 system downtime reporting in an end office.

Consider a population of four systems comprising 1600 terminations distributed as 100, 200, 300, and 1000 terminations per system. The 100- and 1000-termination systems are host systems and each experience one 10-minute total outage during the quarter. The 200- and 300-termination switches are remote systems. The 200-termination system incurred a 20-minute partial outage affecting 50 terminations. The calculation is given below for average switch size = 400 terminations:

Table 4.3 Example: system outage—switching systems.

| Outage length (minutes) | Termi-nations affected | Equivalent Switch factor | Number of Outages | Normalized Frequency Factor* (Number/ Switches) | Normalized Time Factor* (Minutes/ Switches) |
|---|---|---|---|---|---|
| 10 | 100 | 100/400= 0.25 | 1 | 0.250 | 2.5 |
| 10 | 1000 | 1000/400= 2.50 | 1 | 2.500 | 25.0 |
| 20 | 50 | 50/400= 0.125 | 1 | 0.125 | 2.5 |
| Total Quarter | | | M=3 | 2.875 | 30.0 |
| Annualized | | | 12 | 11.50 | 120.0 |

Dividing by the population of 4 Switches, the normalized numbers are:

| Per Switch Per Year | 2.875 Outages (Weighted*) | 30 Minutes |
|---|---|---|
| Per Switch per year | 3.000 Outages (unweighted**) | |

\* Weighted outages indicate average outage frequency for a system of typical size.
\*\* Unweighted outages indicate average outage frequency for any system without regard for size.

Hence, a typical system (or a typical line) will experience 2.875 outages totaling 30.0 minutes in a year based on performance in the current quarter. The metric is independent of whether the outage is total or partial and whether the termination is served by a host or remote system. For systems with incremental recovery, the total duration and frequency can be calculated as the sum of the partial weighted outages.

## THE HARDWARE ONLY METRIC

There is one metric which is applicable to hardware only: Return Rate.

### Return Rate

The circuit pack return rate applies to hardware products only. However, bulk items such as cable, optical fiber, and mechanical hardware are exempted from this measurement.

The purpose of collecting this metric is threefold: 1) To provide a measure of quality of the product initially received by the customer and during subsequent in-service operation; 2) To determine areas that need corrective action or will most likely benefit from continuous improvement activity; and 3) To provide input data needed to calculate equipment lifecycle costs.

Return rates are measured for hardware products as follows:

- Normalized first year percentage returns, and
- Long term annualized percentage returns.

The first year return rate provides an early indication of the in-service performance of both new and existing products. For new products, it provides an indication of the expected performance in the absence of any historical data. For existing products, it provides a means of comparing the expected long-term performance with actual performance when that information becomes available.

The return rate metric is computed for each product category in two parts: (1) Normalized First Year Percentage Returns, and (2) Long-Term Annualized Percentage Returns. The first year return (FYR) rate is rate of returns for 12 months after initial shipment for each product category. The long-term return (LTR) is the rate after the first year (months 13 and later after initial shipment) for each product category.

An example of calculation of return rate is given as follows.

The return rates for January, 1998, are calculated as:

$$\text{Early Return Rate} = \frac{(\text{Returns for units shipped Oct-97 through Jan-98}) \times 12 \times 100}{\text{Shipments for Oct-97 through Dec-97}}$$

$$= \frac{(19 + 21 + 18 + 5) \times 12 \times 100}{9818 + 9787 + 10528}$$

$$= 2.51\%$$

$$\text{First-year Return Rate} = \frac{(\text{Returns for units shipped Jan-97 through Jan-98}) \times 12 \times 100}{\text{Shipments for Jan-97 through Dec-97}}$$

$$= \frac{(11+7+14+10+6+6+9+11+10+19+21+18+5) \times 12 \times 100}{(6263+6436+7244+7275+7396+8263+8833+8954+9368+9818+9787+10528)}$$

$$= 2.44\%$$

$$\text{Long-term Return Rate} = \frac{(\text{Returns from shipments prior to Jan-97}) \times 12 \times 100}{\text{Shipments prior to Jan-97}}$$

$$= \frac{59 \times 12 \times 100}{30000}$$

$$= 2.36\%$$

The return rates for all months in 1998 are plotted in Figure 4.6.

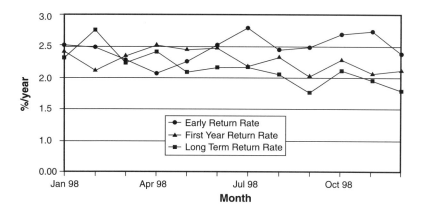

**Figure 4.6** 1998 Return rates.

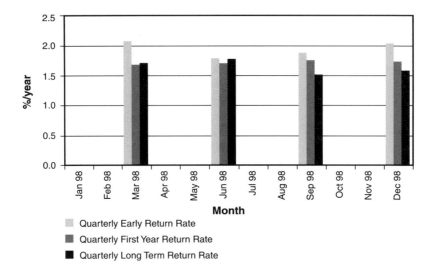

**Figure 4.7** Quarterly return rates.

To calculate the rates for the first quarter (Jan-98 through Mar-98), add the return totals and shipment totals for the three months:

$$1Q98 \text{ Early Return Rate} = \frac{(53 + 60 + 49) \times 12 \times 100}{(30133 + 30959 + 32493)}$$

$$= 2.08\%$$

The quarterly return rates for 1998 are plotted in Figure 4.7.

## SOFTWARE ONLY METRICS

There are three metrics that apply to software only: (1) Software Update Quality, (2) Release Application Aborts, and (3) Corrective Patch and Feature Patch Quality.

### Software Update Quality (SWU)

A software update is a set of changes to a release and is commonly referred as *dot* or *point* release. It differs from a patch in the manner in which software changes are made to a system. A software update replaces existing product with a complete new load. On the other hand, a patch replaces a subset of software in the current load without changing the rest of the load.

The SWU metric tracks the percentage of defective software updates. It is the cumulative percent of defective updates for each of the three latest, most dominant releases per product category. The purpose of tracking this metric is to evaluate the level of defective software updates in order to minimize associated customer risk when implementing software updates. The measurements also show the overall trend of defective software updates by release over a twelve-month period.

The intent of tracking this metric is to ensure that suppliers of software carry out the requisite testing of their software (for example, regression tests or other methods). It has been the experience that software companies, in order to provide a quick solution to a problem in the field, provide an update without undergoing sufficient testing to ensure that the fix does not break something else in the system.

The RQMS[3] contains an alternate method of reporting. Figure 4.8 is an example of a software update metric. The number of software updates is tracked per month for each release. Also, the number of defective software updates is tracked per month for each release. Both monthly and cumulative numbers are currently reported to the MRS, however only the cumulative numbers are used for industry calculations.

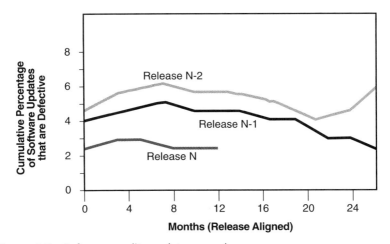

**Figure 4.8** Software quality update example.

## Release Application Abort (RAA)

The release application abort (RAA) metric is the percentage of release applications that abort. It measures the percentage of systems with aborts for each of the last three most dominant releases per product category. This metric is derived from RQMS[3] and RQMS data is an alternative method of reporting the metric.

The purpose of RAA is to minimize the service provider's risk of aborts when applying a software release. The audience includes customer personnel that use the product and/or measure supplier performance. The audience also includes supplier personnel who are involved in the software design/delivery processes as well as those involved in monitoring/managing the vendor quality system.

Table 4.4 is an example of RAA. A vendor is upgrading three active releases (from prior releases to release N, N-1, and N-2). Software upgrades counts and those encountering release application aborts are as shown.

The corresponding release application abort plot is shown in Figure 4.9.

**Figure 4.9** Release application abort metric.

Table 4.4 Release aborts example.

| Month | 1 | 2 | 3 | 4 | 5 | 6 | 7 | 8 | 9 | 10 | 11 | 12 | 13 |
|---|---|---|---|---|---|---|---|---|---|---|---|---|---|
| Number of Upgrades in Month | | | | | | | | | | | | | |
| Release N-2 | 1 | 2 | 5 | 10 | 20 | 40 | 45 | 46 | 45 | 43 | 45 | 30 | 24 |
| Release N-1 | 1 | 3 | 5 | 12 | 20 | 39 | 46 | 51 | 52 | 45 | 48 | 33 | 29 |
| Release N | 1 | 4 | 6 | 14 | 22 | 39 | 45 | 52 | 54 | 50 | 47 | 36 | 30 |
| Cumulative Number of Upgrades to that Release | | | | | | | | | | | | | |
| Release N-2 | 1 | 3 | 8 | 18 | 38 | 78 | 123 | 169 | 214 | 257 | 302 | 332 | 356 |
| Release N-1 | 1 | 4 | 9 | 21 | 41 | 80 | 126 | 177 | 229 | 274 | 322 | 355 | 384 |
| Release N | 1 | 5 | 11 | 25 | 47 | 86 | 131 | 183 | 237 | 287 | 334 | 370 | 400 |
| Number of Upgrades that Encountered Release Application Aborts in Month | | | | | | | | | | | | | |
| Release N-2 | 0 | 0 | 0 | 1 | 0 | 1 | 1 | 1 | 1 | 1 | 1 | 1 | 0 |
| Release N-1 | 0 | 0 | 0 | 1 | 0 | 1 | 1 | 1 | 1 | 1 | 0 | 0 | 1 |
| Release N | 0 | 0 | 0 | 0 | 1 | 0 | 1 | 0 | 1 | 0 | 1 | 0 | 0 |
| Cumulative Number of Upgrades that Encountered Release Application Aborts to that Release | | | | | | | | | | | | | |
| Release N-2 | 0 | 0 | 0 | 1 | 1 | 2 | 3 | 4 | 5 | 6 | 7 | 8 | 8 |
| Release N-1 | 0 | 0 | 0 | 1 | 1 | 2 | 3 | 4 | 5 | 6 | 6 | 6 | 7 |
| Release N | 0 | 0 | 0 | 0 | 1 | 1 | 2 | 2 | 3 | 3 | 4 | 4 | 4 |
| Release Application Aborts Metric (Cumulative %) | | | | | | | | | | | | | |
| Release N-2 | 0.00 | 0.00 | 0.00 | 5.56 | 2.63 | 2.56 | 2.44 | 2.37 | 2.34 | 2.33 | 2.32 | 2.41 | 2.25 |
| Release N-1 | 0.00 | 0.00 | 0.00 | 4.76 | 2.44 | 2.50 | 2.38 | 2.26 | 2.18 | 2.19 | 1.86 | 1.69 | 1.82 |
| Release N | 0.00 | 0.00 | 0.00 | 0.00 | 2.13 | 1.16 | 1.53 | 1.09 | 1.27 | 1.05 | 1.20 | 1.08 | 1.00 |

## Corrective Patch Quality (CPQ) and Feature Patch Quality (FPQ)

A feature patch adds functionality, while a corrective patch fixes a known problem. For each of these types of patches, errors can occur.

Corrective patch quality is the percentage of official corrective patches that are determined to be defective. Feature patch quality is defined as the percentage of official feature patches that are determined to be defective. The purpose of tracking these metrics is to evaluate the percentage of defective official patches in order to minimize service provider risk of failure.

Both corrective patch and feature patch quality metrics are computed for each of the latest three most dominant releases. They are applicable to software products only. These metrics are adapted from RQMS[3] and RQMS data is a valid reporting alternative. The Metrics Handbook[7] describes in detail the counting rules, calculations and formulas, reporting format, etc. An example of defective corrective patch metric is shown in Table 4.5.

The corresponding corrective patch quality metric plot is shown in Figure 4.10.

**Figure 4.10** Corrective patch quality example.

**Table 4.5** Example of corrective patch quality.

| Month | 1 | 2 | 3 | 4 | 5 | 6 | 7 | 8 | 9 | 10 | 11 | 12 | 13 |
|---|---|---|---|---|---|---|---|---|---|---|---|---|---|
| **Number of Corrective Patches Issued in Month** | | | | | | | | | | | | | |
| Release N-2 | 60 | 55 | 50 | 47 | 42 | 35 | 35 | 31 | 32 | 30 | 29 | 27 | 25 |
| Release N-1 | 55 | 55 | 50 | 40 | 36 | 32 | 34 | 36 | 33 | 32 | 26 | 24 | 24 |
| Release N | 52 | 53 | 48 | 35 | 34 | 34 | 32 | 30 | 28 | 30 | 25 | 24 | 22 |
| **Number of Defective Corrective Patches Identified in Month** | | | | | | | | | | | | | |
| Release N-2 | 1 | 0 | 0 | 2 | 0 | 0 | 0 | 1 | 0 | 1 | 1 | 0 | 0 |
| Release N-1 | 1 | 0 | 0 | 1 | 0 | 0 | 1 | 0 | 0 | 0 | 1 | 0 | 1 |
| Release N | 0 | 1 | 0 | 0 | 0 | 1 | 0 | 0 | 0 | 0 | 0 | 1 | 0 |
| **Defective Corrective Patch Metric (6 Month Rolling Window— % Defective Corrective Patches)** | | | | | | | | | | | | | |
| Release N-2 | 1.67 | 0.87 | 0.61 | 1.42 | 1.18 | 1.04 | 0.76 | 1.25 | 1.35 | 0.98 | 1.56 | 1.63 | 1.72 |
| Release N-1 | 1.82 | 0.91 | 0.63 | 1.00 | 0.85 | 0.75 | 0.81 | 0.88 | 0.95 | 0.49 | 1.04 | 1.08 | 1.14 |
| Release N | 0.00 | 0.95 | 0.65 | 0.53 | 0.45 | 0.78 | 0.85 | 0.47 | 0.52 | 0.53 | 0.56 | 0.59 | 0.63 |

A supplier has three active releases (N, N-1, and N-2). The table above shows how the corrective patch distribution and bad corrective patch counts have been calculated.

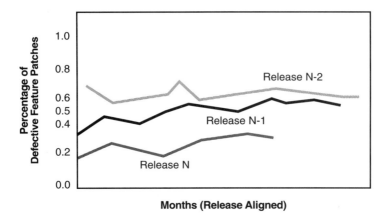

**Figure 4.11** Example: feature patch quality.

The plot shown in Figure 4.11 is an example of defective feature patch quality.

# THE SERVICE ONLY METRIC

There is only one metric which is applicable to service side: Service Quality.

## Service Quality

The service quality metric is applicable only to services. It can be applied to any type of service provided by a supplier. Some of the metrics measured are:

- Percent conforming installation audits;
- Percent successful maintenance visits without callbacks;
- Percent successful repairs;
- Percent conforming call center calls resolved within agreed upon time; and
- Percent support service transactions without a defect.

The purpose of this metric is to provide quality measurement information for establishing the evaluation and continuous improvement of the service.

The following are examples of the service quality metric:

A. *Installation Example:*

1. Data Collected and Results.

**Example Source Data for Installation SQ**

|  | January | February | March | April |
|---|---|---|---|---|
| # of Nonconforming Installation Audits | 5 | 1 | 0 | 6 |
| Total Number of Installation Audits | 100 | 50 | 75 | 80 |
| Service Quality Metric | 95% | 98% | 100% | 92.5% |

2. Computation for the Month of January:
$$100 \times (1-5/100) = 95\%$$

B. *Engineering Example:*

1. Data Collected and Results.

**Example Source Data for Engineering SQ**

|  | January | February | March | April |
|---|---|---|---|---|
| # of Engineering Corrections | 5 | 2 | 0 | 4 |
| Total Number of Billable Engineering Staff Hours (100s) | 10.00 | 5.00 | 7.50 | 3.00 |
| Service Quality Ratio Metric | .5 | .4 | 0 | 1.333 |

2. Computation for the Month of January:
$$5/10.00 = .50 \text{ defects per 100 staff hours.}$$

C. *Maintenance Example:*

1. Data Collected and Results.

**Example Source Data for Maintenance SQ**

|  | January | February | March | April |
|---|---|---|---|---|
| # of Callbacks | 2 | 0 | 1 | 4 |
| # of Maintenance Visits | 30 | 20 | 75 | 120 |
| Quality Service Conformance Metric | 93.3% | 100% | 98.6% | 96.7% |

2. Computation for the Month of January:
$$100 \times (1-2/30) = 93.3\%$$

D. *Call Center Example:*

1. Data Collected and Results

**Example Source Data for Call Center SQ**

|  | January | February | March | April |
|---|---|---|---|---|
| # of Call Resolutions That Exceeded the Specified Time Allotment | 15 | 40 | 10 | 4 |
| Total # of Calls Which Came into the Call Center | 2000 | 5000 | 2750 | 3000 |
| Quality Service Conformance Metric | 99.25% | 99.2% | 99.6% | 99.8% |

2. Computations for the Month of January:
$$100 \times (1-15/2000) = 99.25\%$$

E. *Support Services Example:*

This example references a cable locator service with a defined defect as a cut cable due to incorrect identification.

1. Data Collected and Results

**Example Source Data for Support Services SQ**

|  | January | February | March | April |
|---|---|---|---|---|
| Cut Cables (*# of defects*) | 5 | 2 | 0 | 4 |
| # of Cables Identified (*# of opportunities for defects*) | 1000 | 500 | 750 | 300 |
| Quality Service Conformance Metric | 99.5% | 99.6% | 100% | 98.7% |

2. Computation for the Month of January:
$$100 \times (1-5/1000) = 99.5\%$$

## METRICS SUBMISSION AND ANALYSIS

Obviously, for all of this to work, each party involved in the process, that is, the supplier, their customers, the registrars, and the database administrator, must all perform certain tasks and meet certain responsibilities.[8]

The metrics work group chartered a subgroup to create specifications of the metrics repository system, the registration repository system, the confidentiality and security aspects of data. The metrics subgroup prepared an elaborate specification for such a system. The next chapter describes these details.

## END NOTES

1. *Criteria for Performance Excellence* (Gaithersburg, MD: Baldrige National Quality Program, 2000).
2. The European Foundation for Quality Management (Brussels, Belgium, 1999).
3. GR-929-CORE, *Reliability and Quality Measurement for Telecommunications Systems (RQMS) (*Morristown, NJ: Telcordia Technologies, Issue 4, December 1998).
4. GR-1315-CORE, *In-Process Quality Metrics (IPQM) (*Morristown, NJ: Telcordia Technologies, June 1995).
5. GR-1323-CORE, *Supplier Data—Comprehensive Generic Requirements* (Morristown, NJ: Telcordia Technologies, Issue 1, December 1995).
6. V. Basili and D.M. Weiss. "A Methodology for Collecting Valid Software Engineering Data." *IEEE Transactions on Software Engineering* SE-10, no. 6 (November 1984): 728–738.
7. *TL 9000 Quality System Metrics*, Book Two, Release 2.5 (The QuEST Forum, 1999).
8. Galen Aycock, Jean-Normand Drouin, and Thomas Yohe. "TL 9000 Performance Metrics to Drive Improvement." *ASQ Quality Progress* 32, no. 7 (July 1999).

# 5

# TL 9000 Metrics Data Management and Analysis

## METRICS REPOSITORY SYSTEM

The University of Texas at Dallas (UTD) is responsible for the Metrics Repository System (MRS). UTD was selected by the QuEST Forum from a pool of very competent contenders. The most sensitive aspect of the MRS is data security. Due to intensive competition between suppliers, they are not comfortable sharing data with each other.

In order to address these concerns, the metrics work group chartered a subteam to define requirements on the MRS. The team prepared a detailed system specification. The main items in the list include:

- MRS specifications describe type of technology, database, data entry, security, and confidentiality, data submission, data validation, etc.

- Security aspects include:
  - Key management: access control
  - Metrics data submission: encrypted keys; who issues encrypted keys; who uses the keys; etc.
  - Metrics data storage: data transfer and storage; temporary and permanent storage; etc.

The metrics work group posed the following two primary requirements on the MRS. UTD ensured that they satisfied these requirements:

1. Measurement data must remain secure throughout the process, from its origin through analysis and posting; and

2. No single person should be able to compromise the security, confidentiality, or reliability of submitted measurement data.

UTD complied with these requirements effectively. The data security has three important elements, which are:

1. Confidentiality of source and data;
2. Integrity of data; and
3. Availability of processed data.

## Confidentiality of Source and Data

The QuEST Forum appointed a subteam to define requirements on the MRS. Confidentiality of source and data is the single most important concern of suppliers. UTD worked with the subteam to alleviate such concerns. UTD incorporated two key design elements, which are:

1. Separation of administrative system from the data system; and
2. Encryption.

The administrative system, called the registration repository system (RRS), is managed by ASQ. When suppliers decide to apply for TL 9000 registration, they notify ASQ of their intention. ASQ requires suppliers to define the scope (hardware, software, service, or any combination) and product categories. ASQ then assigns reporting IDs for each product category. The supplier submits the data to UTD using the reporting IDs. UTD does not have knowledge of who is submitting the data, nor does ASQ have knowledge of the data content. This is the essence of the "double blind" system. Later in the process, ASQ issues a Data Submission Confirmation to signify the supplier's successful submission of data to UTD.

The TL 9000 metrics database (MRS) is managed by UTD. The supplier encrypts the data prior to submission, using the data submission software provided by UTD. The supplier chooses the encryption key, which is 128 bits long. UTD receives the encrypted data from the supplier, but they do not know the owner of the data. This is because the data are transmitted from the incoming server directly into the repository computer. UTD checks for a valid reporting ID with data. They also perform a sanity check on the data. If the data are valid, they provide a qualitative data confirmation report to ASQ. The MRS and RRS are tied together through the reporting ID in the "double blind" system.

UTD developed a secure process to ensure that the database is backed up and recoverable in the event of natural disasters. UTD also instituted methods to protect the database from infection by software viruses.

## Encryption Process

The encryption process requires that the submitter choose the encryption key that is 128 bits long (16 ASCII characters). The submitter checks and encrypts their data. The submitter is then required to encrypt their own encryption key using public/private encryption of 1024 bits. The encrypted encryption key along with the encrypted data is then sent to UTD using one of the following options:

- Anonymous submittal of encrypted data
  - Transmission of data via internet; or
  - Mailing of data via courier (FedEx, etc.).
- Nonanonymous submittals of encrypted data
  - Direct transmission of data via the internet; or
  - Direct mailing of data via courier.

At the other end in UTD, the reporting ID is the only identifying element. UTD does not have the knowledge of who owns which reporting ID. Only data with verifiable reporting IDs are accepted by the database program. Also, only encrypted verifiable data from the supplier enter the database. No other data will be accepted by the program. The input data are not seen by anyone at ASQ or UTD.

In summary, the encryption process ensures the following:

- Data is always encrypted; and
- Unencrypted input data are never shown or seen.

## Data Integrity

UTD has made elaborate arrangements to house the MRS system. They have installed stand-alone computers in a locked room. Both encrypted data and computers reside in a locked room. The computer room is under constant video surveillance. The videotapes are regularly viewed for illegal entry. The entry logs are verified twice a week. Nobody has unauthorized access to the computers. All software is strictly controlled.

The system consists of triple redundant computers. The encrypted data are backed up. The storage computers receiving encrypted data are not connected to any network, so hackers can never access the system. The computer room power is protected by double back-up power supplies. Also, the staff handling data is specially trained on the MRS process.

The MRS system complies with BS 7799,[1] the international standard on Information Security. BS 7799, first published in February 1995, is a comprehensive set of controls comprising best practices in information security. BS 7799 is intended to serve as a single reference point for identifying a range of controls needed for most situations where information systems are used in industry and commerce, and to be used by large, medium, and small organizations. The current document was significantly revised and improved in May 1999.

As part of the BS 7799 assessment, UTD's MRS system was audited by the British Standards Institution (BSI) assessors. BS 7799 covers all aspects of data security through 130 auditable controls. UTD achieved BS 7799 certification in January 2000. The metrics repository system at UTD is the first BS 7799 certified secure system in the United States. They will be audited semi-annually by BSI.

ASQ is in the process of obtaining BS 7799 certification for the RRS system.

## METRICS DATA STORAGE

Encrypted supplier data submitted using the internet are instantaneously transferred to an intermediate computer not accessible via the internet. The data are then physically moved to temporary storage at the stand-alone, highly secure computer system without decryption. Data submitted via anonymous FedEx are stored only on the stand-alone computer system without decryption.

Supplier data are stored in the secure database when authorized. Storing data in the secure database can only be activated by two persons, for example, two keys must be entered by two key holders for authorization. A two person password is required for database access to ensure that no single person can compromise (decrypt) the data.

The supplier encryption keys are recovered first to decrypt the data. The data are then encrypted using the UTD key [$(N, K)$ scheme] and stored. This is a password security scheme where $N$ is the total number of passwords and it takes any $K$ passwords to achieve access. Each password is 16 characters long.

UTD sends a data confirmation report to a supplier via ASQ. The report does not contain any data. A sample data confirmation report is shown in Table 5.1 for the number of problem reports (NPR) metric. MetricID is a key word. NPR is the metric name (number of problem reports).

Table 5.1  Sample data confirmation report.

| MetricID | NPR | CustomerBase | Total |
|---|---|---|---|
| NPRS | Ok | Ok | Ok |
| NPRA | Ok | Ok | Ok |
| Np1 | OutOfRange | Ok | Missing |
| Np2 | Ok | Ok | Ok |
| Np3 | Ok | Ok | Ok |
| Np4 | NotApplicable | NotApplicable | NotApplicable |

CustomerBase is a key word referring to the number of customers who are QuEST Forum members. Total represents all customers. NPRS is the normalization factor. NPRA is the annualization factor. Np1, Np2, Np3, and Np4 refer to the number of critical, major, minor, and service problems respectively. Each item in a column represents a value for one month. The software does a sanity check on the values and determines if the data points are within the range (Ok); Missing; OutOfRange; or NotApplicable. The input software advises range values before the data are sent in.

## METRICS DATA FLOW

The metrics system involves data collection, analysis, and reporting activities. Each of these activities consists of interaction between the following entities:

- Supplier;
- Customer;
- University of Texas at Dallas (UTD);
- American Society for Quality (ASQ); and
- Registrar.

The "double blind" system was set up to insure that the data are secure and not available to competitors and other unauthorized persons. For the five entities to work together efficiently in the double blind system, it requires extensive communication and passage of data and information. The five entities must interact with each other at various times. Figure 5.1

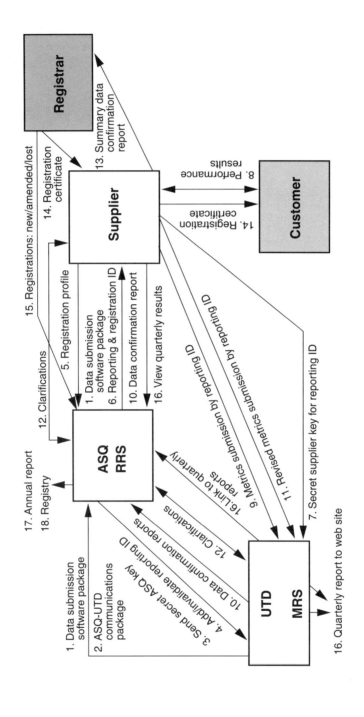

**Figure 5.1** Interactions in the metrics implementation process.

# TL 9000 Metrics Data Management and Analysis 113

shows the overall system, and Figures 5.2 through 5.6 show the paired and triplet interactions.

There are 18 interactions shown in Figure 5.1. Table 5.2 lists these interactions.

We will describe the paired and triplet interactions in the next sections of this chapter. Numbers shown in parentheses in those sections indicate the part of the process covered.

**Table 5.2** Interactions during metrics data flow.

| Interaction | From | To |
|---|---|---|
| 1. Data submission software package | UTD<br>ASQ | ASQ<br>Supplier |
| 2. ASQ-UTD communications package | UTD | ASQ |
| 3. Send secret ASQ key | ASQ | UTD |
| 4. Add/invalidate reporting ID | ASQ | UTD |
| 5. Registration profile | Supplier | ASQ |
| 6. Reporting & registration IDs and the password | ASQ | Supplier |
| 7. Secret supplier key for reporting ID | Supplier | UTD |
| 8. Performance results | Supplier<br>Customer | Customer<br>Supplier |
| 9. Metrics submission by reporting ID | Supplier | UTD |
| 10. Data confirmation reports | UTD<br>ASQ | ASQ<br>Supplier |
| 11. Revised metrics submission by reporting ID | Supplier | UTD |
| 12. Clarifications | UTD<br>ASQ<br>Supplier<br>ASQ | ASQ<br>Supplier<br>ASQ<br>UTD |
| 13. Summary data confirmation report | ASQ<br>Supplier | Supplier<br>Registrar |
| 14. Registration certificate | Registrar<br>Supplier | Supplier<br>Customer |
| 15. Registrations: new, amended, lost | Registrar | ASQ |
| 16. Quality report to web site | UTD<br>ASQ | ASQ<br>QuEST Forum web site |
| 17. Annual report | ASQ | QuEST Forum web site |
| 18. Registry | ASQ | QuEST Forum web site |

## Interaction Between Supplier and Customer

The interaction between customer and supplier is critical to the metrics system. There are many metrics which require customers to provide the data to the suppliers. Also, TL 9000 Requirements mandate stronger customer and supplier relationships.

For example, computation of number of problem reports (NPR) requires that only customer-originated problem reports be counted. So, it is imperative that customers have processes in place to capture problem reports, and then provide necessary data to suppliers to generate measurement.

The interaction starts when a customer inquires, requests, or contracts with suppliers for TL 9000 registration. The supplier notifies the customer of its plans for TL 9000 registration and requests data to complete the metrics calculations. The supplier and customer together need to establish a process to receive and validate the data on a regular basis (8). The supplier validates the data and requests clarifications when necessary.

The supplier completes the required TL 9000 metrics computations. When the supplier receives certification, it notifies the customer (14). The supplier may meet with customers to provide customer specific reports with industry norms received from UTD. The supplier and customer also may jointly establish improvement teams. The interaction between customer and supplier is depicted in Figure 5.2.

**Figure 5.2** Interactions between customer and supplier.
*The dashed arrow indicates the flow of information on request.*

## Interaction Between Supplier and ASQ

The relationship between the supplier and ASQ begins when the supplier establishes an account with ASQ and provides a registration profile (5). The information provided by the supplier includes:

- Applicable metrics;
- Determination of which will be TL 9000 or RQMS alternative; and
- Scope of the customer base for metrics.

ASQ provides the supplier with the following:

- Access to registration profile with registration ID and password (6);
- Reporting ID for each product category (6);
- Data Submission Software (DSS) package (1); and
- Access to quality results—for QuEST members only (16).

Also, ASQ interfaces with the supplier to obtain clarification (12) of any data concerns. ASQ reroutes the UTD data confirmation report for each reporting ID (10) and mails a summary data confirmation report for each registration. A pictorial representation of this interaction is shown below in Figure 5.3.

**Figure 5.3** Interaction between supplier and ASQ.

## Interaction Between Supplier and UTD

The supplier uses data submission software (DSS) (1) to:

- Transmit secret supplier key to UTD (7);
- Validate metrics data;

**116** Chapter Five

- Encrypt metrics data and transmit them to UTD (9); and
- If the transmission is successful, UTD sends a confirmation report back to the supplier via ASQ (10).

UTD in turn does the following:

- Validate metrics data;
- Identify missing or out-of-range data; and
- If needed, ask supplier (through ASQ) to send revised data or other clarifications (12).

Also, the supplier provides revised metrics to UTD in case of an error or transmission failure (11). If the transmission fails, the supplier works through ASQ to resolve the issues (12). When data has been successfully accepted by UTD, a data confirmation report is sent to the supplier (10).

These interactions are summarized in Figure 5.4.

**Figure 5.4** Interaction between supplier and UTD.

## Interaction Among Supplier, ASQ, and UTD When Data Transmission Fails

If the transmission fails, the supplier works through ASQ to resolve the issues (12). UTD in turn does the following:

- Validates metrics data;
- Identifies missing or out-of-range data;
- If needed, asks supplier (through ASQ) to send revised data (12).

# TL 9000 Metrics Data Management and Analysis    117

**Figure 5.5** Interaction among supplier, ASQ, and UTD when transmission fails.

The supplier resubmits metric data to UTD (11). If data errors continue, UTD consults with ASQ. When data have successfully been accepted by UTD a data confirmation report is sent to the supplier (10). The three-way interactions are shown in Figure 5.5.

## Interaction Among Supplier, Registrar, and ASQ

The interaction among the supplier, registrar, and ASQ is quite extensive. The supplier sends ASQ the registration profile (5). After receiving the registration ID (6) from ASQ, the supplier submits to assessment by the registrar. The registrar assesses the supplier's compliance to:

- TL 9000 requirements, and
- TL 9000 metrics.

The registrar also assesses the supplier's data validity and integrity processes. They also verify that metrics are reported for old and new products and review the summary confirmation report (13) as a quality record.

The registrar writes nonconformances if necessary and ensures that the supplier provides an action plan to address the deficiencies. The supplier is also required to demonstrate to the auditor that the nonconformances against their system are resolved in a timely manner. After a satisfactory assessment, the registrar grants registration (14) to the supplier and notifies ASQ of registration (15). The interactions between these three entities are captured in Figure 5.6.

**Figure 5.6** Interaction among supplier, ASQ, and registrar.

### Interaction Between ASQ and UTD

The interaction between ASQ and UTD is very crucial for the MRS system. UTD has defined the communication process between UTD and ASQ. UTD provides the communication package to ASQ (2). ASQ uses the communication package to provide the following to UTD:

- Secret ASQ key (3), and
- Valid reporting ID (4).

In addition, ASQ receives the data submission package from UTD (1) which they make available for suppliers (with valid key) through the web site. UTD and ASQ work jointly to resolve any chronic issues with supplier data (12). UTD also compiles quarterly reports and provides ASQ with web link (16). ASQ provides gatekeeper functions to access quarterly data at the web site (16). Later, UTD sends ASQ a data confirmation report (10). See Figure 5.7.

### ASQ's Additional Responsibilities

In addition to the interaction with other entities, ASQ also is responsible for publishing the annual report (17) and the registry (18) on the QuEST Forum web site.

# THE GOALS AND EXPECTATIONS OF TL 9000 METRICS

There were five major goals defined for the QuEST Forum:

1. To drive continuous improvement in products, services, and processes;
2. To develop standard metrics to drive product quality improvement;

## TL 9000 Metrics Data Management and Analysis

**Figure 5.7** Interaction between ASQ and UTD.

3. To implement an industry standard assessment process which would reduce the multiplicity of conflicting programs;

4. To foster systems that will protect the integrity of telecommunications products, services, and networks; and

5. To develop requirements that will help organizations more accurately assess the implementation of their quality management system.

There are a number of industry benefits that the QuEST Forum membership expects from the development of TL 9000. These include:

- Improved service to end users;

- Enhanced customer-supplier relationships (note: QuEST Forum has been a cooperative effort among customers and suppliers);

- Uniform performance and cost-based metrics to use as benchmarks for improving product and service quality;

- Enhanced supply chain management, including second and third tier suppliers; and

- The creation of a platform for industry improvement initiatives.

Specific expectations for Forum member products and services include:

- Cycle-time reduction;
- Improved on-time delivery;
- Reduced lifecycle costs;
- Increased profitability and market share;
- Superior products; and
- Defect reduction.

## The Use of Metrics to Improve Product and Service Quality

The unique difference from current ISO 9000 practice is the use of metrics. These measurements are used to determine the quality level of telecommunications product and services. Examples are circuit pack return rate, number of problem reports, and software update quality. It is expected that extending ISO 9000 in this direction will result in an industry drive to improve overall quality. Certainly, when suppliers know where their products stand relative to their competitors, and their customers know the quality level of "best in class," there will be a drive by all to improve.

The QuEST Forum envisions multiple uses for the metrics. For example, they can be used to enhance customer-supplier communication that will be used to prioritize and solve the most cost-effective problems. Metrics can also be used to bridge the gap between quality issues and business results. And, they can be used to quantify the customers' views of their suppliers' quality. Some service providers have indicated that they will use the metrics to develop report cards for their suppliers.

The Forum has retained the University of Texas at Dallas (UTD) to gather the data and develop statistics for each product class. The statistics will identify the mean (or median) standard deviation (or range) and "best in class." The data will be published on the QuEST Forum web site (www.questforum.org).

A sample plot showing the three measurements for on-time delivery metric is shown in Figure 5.8.

The key to confidentiality is that no one knows where a particular supplier's data falls on this chart. Each supplier alone knows where they stand on the plot.

Also, output from the metrics database consists of monthly or quarterly statistical summary reports derived from the TL 9000 metrics repository for each metric by product category as posted on the Quest Forum web site.

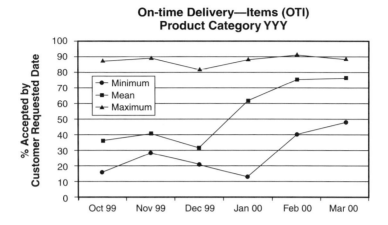

**Figure 5.8** Sample on-time delivery metric.

These reports are posted on the web site and distributed in hardcopy form by ASQ as appropriate. However, the statistics for a product category are computed and posted only if it can be assured that the statistical results will not in any way compromise supplier anonymity. There must be at least five sets of data from three companies before computations can be made.

UTD was selected because of its long-standing role in telecommunications engineering education. It's expected that UTD will provide insight into benchmarking industry quality problems, identifying quality trends, and identifying possible solutions to quality problems.

## Metrics Computation

A major issue during the development of these metrics is the identification of "normalization factors" which will allow comparison of similar products with different characteristics. For example, for circuit switching, the normalization factor for return rate is returns/10,000 terminations/year. Thus products with a large number of terminations per circuit pack are measured fairly against those with fewer terminations per circuit pack.

Another issue is a result of a requirement of some service providers that the suppliers provide "Reliability and Quality Measurements for Telecommunications Systems" (RQMS) metrics defined in the Telcordia Technologies document GR-929-CORE.[2] In this case the supplier may provide the RQMS metrics in place of the TL 9000 metrics.

Other metrics and indicators are under consideration, but have not been approved. Indicators are a separate category of measures that are used to "flag potential cost, schedule, productivity, and quality issues."[3]

## Product and Service Categories

Along with the definition of metrics, it was an enormous task to identify products for which the metrics would be applicable. A subteam was chartered to create a list of such products. The team spent quite some time doing research. Their deliverable was the product category table.[4] The table will include all the products in the telecommunications industry that will fall under the metric umbrella. It will be a true challenge to keep the product category table current with the frequent introduction of new products.

In the beginning, the product category table was being updated almost on a monthly basis. Finally, the Forum decided to maintain the latest version of the table on their web site (www.questforum.org). The list continues to increase as more and more products are introduced.

The following is a partial list of product and service categories defined for the metrics implementation (see Appendix A in Book Two):[4]

- *Switching*: Equipment for the physical or virtual interconnection of communications channels in response to a signaling system. The switching category is broadly defined to include packet or circuit switched architectures.

- *Signaling*: Equipment for the provisioning of signaling, that is, "[network] states [that are] applied to operate and control the component groups of a telecommunications circuit to cause it to perform its intended function. . . . There are five basic categories of signals: . . . supervisory, information, address, control, and alerting . . . includes [only] signaling products that function within the telecommunications network. . . ."

- *Transmission*: Equipment for the connection of the switched and interoffice networks with individual customers. An integral part of the distribution network is the loop, which connects the customer to the central office (CO), thus providing access to the interoffice network.

- *Operations and maintenance*: Equipment, systems, and services for the management, upkeep, diagnosis, and repair of the communications network.

- *Common systems*: Any of a variety of specialized generic, shared equipment to support network elements. Common systems include power systems and network equipment-building systems (NEBS) that provide space and environmental support for [the] network.[5]

- *Customer premises*: Equipment installed beyond the network demarcation point. Although commonly installed on the subscriber's premises, equipment with essentially identical functionality installed in the service provider's facility may also be classified as customer premises equipment.

- *Services (installation, engineering, maintenance, repair, call center, and support)*: Results generated by activities at the interface between the supplier and the customer and by supplier internal activities to meet customer needs. Note that:
  - The supplier or customer may be represented at the interface by personnel or equipment;
  - Customer activities at the interface with the supplier may be essential to the service delivery;
  - Delivery or use of tangible product may form part of the service delivery; and
  - A service may be linked with the manufacture and supply of tangible product.

- *Components and Subassemblies (includes component suppliers, contract manufacturers, and OEM suppliers)*: Individual components or assemblies provided for use in telecommunications systems excluding those already covered by a specific product category in another product family. These items would typically be used by other suppliers and not sold directly to service providers except as replacement parts.

## Training

Since the metric requirement is unique to TL 9000, the QuEST Forum arranged training for suppliers, registrars, and trainers. The Forum appointed two training organizations, EXCEL Partnership and STAT-A-MATRIX. They worked with the two training organizations and UTD to develop a number of metrics-related courses. Some of the courses are:

- *Metrics Collection and Analysis*—focused on the planning aspect associated with design and implementation of metrics collection and reporting systems;

- *QuEST Forum Sanctioned TL 9000 Metrics Data Submission*—focused on how to perform data submissions, how the security system was installed for MRS, and how to use data for internal improvements;

- *Metrics Implementation*—designed to help suppliers understand the importance of metrics and help them design a system for TL 9000 metrics collection, analysis, reporting, and submission.

The Forum was actively involved during the course development to ensure that a correct and consistent message is given out regarding TL 9000 metrics. The pilot organizations took early versions of the courses as part of the overall pilot program. They provided valuable input to the course developers, which was later used to improve the courses.

## THE QUEST FORUM'S GLOBAL APPROACH

The Forum has also developed a cooperative relationship with the EIRUS organization, a telecommunication metrics user group. EIRUS uses two sets of metrics based on Telcordia Technologies requirements: (1) European In-Process Quality Metrics (E-IPQM),[6] and (2) European Reliability and Quality Measurements for Telecommunications Systems (E-RQMS).[6] Joint working groups have been established to develop alignment of the EIRUS and QuEST Forum metrics.

## END NOTES

1. BS 7799: 1999, *Information Security Management, Part I: Code of Practice for Information Security Management and Part II: Specification for Information Security Management* (London: The British Standards Institute, 1999).
2. GR-929-CORE, *Reliability and Quality Measurements for Telecommunications Systems (RQMS)* (Morristown, NJ: Telcordia Technologies, Issue 4, December 1998).
3. Henry Malec. "TL 9000 Database Repository and Metrics." *The Informed Outlook* 4, no. 6 (June 1999): 4.
4. *TL 9000 Quality System Metrics*, Book Two, Release 2.5 (The QuEST Forum, 1999).
5. GR-63-CORE, *Network Equipment-Building System (NEBS) Requirements: Physical Protection* (Morristown, NJ: Telcordia Technologies, Issue 1, October 1995).
6. "Reliability Engineering." *European Quality Metrics: E-IPQM and E-RQMS*, Deliverable 5, EURESCOM P307 (November 1995). EURSECOM GmbH, Scloss Wolfsbrunnenweg 35, 69118 Heidelberg, Germany, www.eurescom.de/.

# 6

# TL 9000 Case Studies: An Innovative Way of Gathering Feedback Information

## INTRODUCTION

In February 1999, the QuEST Forum took an extraordinary step: it launched a pilot program initiative prior to general publication of the handbooks. Eleven telecommunications suppliers and two service providers (TSPs) were enlisted to implement TL 9000 within one or more facilities and with the goal of obtaining registration. The pilot program was started before publication of the first editions of *TL 9000 Quality System Requirements*[1] in May 1999 and *TL 9000 Quality System Metrics*[2] in November 1999.

The pilot program had multiple feedback goals. First, it was intended to obtain information concerning implementation issues. The pilot organizations held monthly meetings to provide information to the pilot participants. Secondly, it was a training ground for the accreditation agencies, the registrars, the training organizations, and the data gatherers. Finally, it was a means of recognizing the organizations participating in the program.

*Author's note*: This chapter is an adaptation of an article written by Jim Mroz that appeared in *The Informed Outlook*, April 2000.* The author gathered the survey data and collaborated with Jim in the development of the article. I want to thank Jim and the publishers, Robin and Jim Gildersleeve, for working with us to develop the material in this chapter.

---

*"TL 9000 Pilot Program Case Studies Enhance Focus," *The Informed Outlook* 5, number 4 (April 2000): 1, 27–33.

The pilot program is an example of the proactive efforts the QuEST Forum has made to ensure that the TL 9000 system actually raises the baseline, produces benefits for its users, and measures QMS results to accelerate company- and sectorwide improvements. The pilot initiative provided the QuEST Forum with feedback on TL 9000 that will be used in making future revisions to the handbooks. The implementation and registration efforts concluded with recognition at a QuEST Forum meeting in Dallas in January 2000. Sixteen organizations from 11 companies were honored. As an example of the extraordinary cooperation of companies in the QuEST Forum, the pilot organizations agreed not to publicize their registrations until the January meeting.

The QuEST Forum's working groups used project planning techniques similar to those used by ISO Technical Committee (TC) 176, when it revised the ISO 9000 series. However, the QuEST Forum's use of a pilot program during the development of its sector-specific requirements and metrics demonstrated a unique approach to "standards development." This is in contrast to the validation program used by ISO/TC 176 to gain experience in applying the revised ISO 9001. In that program, the feedback to TC 176 was based on a pseudo-implementation where the participants did a gap analysis, estimated cost and time to implement, and answered an extensive questionnaire. But the participants were not required to implement ANSI/ISO/ASQ Q9001–2000.

## PILOT PARTICIPANTS

Eleven companies participated in the pilot program. The participants included hardware, software, hardware and software combination, and service organizations. The types of products ranged from wireless transmission equipment, circuit switching equipment, and transmission equipment to fiber optic cable, software products, and installation and engineering services. The organizations were located in the United States, Canada, England, Poland and the Netherlands. Table 6.1 contains a list of the pilot participants.

## CASE STUDY SURVEYS

As part of the effort to gather information for the case studies, the Forum sent surveys to the pilot program participants—both companies that have obtained TL 9000 registration and registrars that conducted

**Table 6.1** Pilot participants.

| Hardware (8) | Software (3) | Hardware & Software (4) | Service (1) |
|---|---|---|---|
| Adtran | Motorola | Fujitsu Network Communications | Nortel Networks |
| NEC America | Nortel Networks | Lucent Technologies | |
| Nortel Networks (2 organizations) | SBC—California | Nortel Networks | |
| Pirelli Cables & Systems | | Marconi Access | |
| Siecor (2 Organizations) | | | |
| Tellabs | | | |

the assessments—to evaluate implementation and registration experiences. Nine of the eleven TL 9000 registered pilot organizations and five of seven registrars completed and returned the survey. In this chapter we will look at the case study survey results and point out guidance on the general trends detected among the pilot program participants. Keep in mind that continual improvement and customer satisfaction monitoring are requirements of ANSI/ISO/ASQ Q9001–2000. The experiences of the pilot ISO registered companies with a new layer of QMS requirements may identify opportunities for other organizations, even those not in the telecommunications industry.

## The Case Study Survey Results

The QuEST Forum case study survey is intended to assess the pilot program results and gain feedback on the experience of organizations that have implemented TL 9000. The names of the respondents and any references to the organizations discussed below have been intentionally omitted to provide a look at the common experiences and unique outcomes without focusing on a particular company.

The case study survey was distributed electronically to the pilot program participants in January 2000 and later in the year to their registrars. The pilot survey was grouped into 16 categories and consisted of 63 questions. The 9 respondents represent 14 of the 16 registrations from the pilot program. The registrar survey questions were grouped into 9 categories consisting of 26 questions.

## Pilot Survey Questions

General questions (3*) (for example, facility products/services, type of registration, value of registration)

1. Beginning the Journey: Getting Commitments (3)
2. Setting the Scope (3)
3. Selecting the Registrar (2)
4. Assessing the Status (gap analysis) (4)
5. Documenting/Developing Processes (2)
6. Implementing the Process (5)
7. Training (4)
8. Internal Audits (3)
9. Pre-assessment (4)
10. Corrective Action from the Internal Audit and/or the Pre-assessment (2)
11. The Certification Audit (5)
12. Reporting TL 9000 Metrics (5)
13. Estimate of Effort (7)
14. Estimate the Value Added by Implementing TL 9000 (4)
15. Review of the Approach (7)

*Numbers in parentheses indicate the number of questions in each category.

## Registrar Survey Questions

General questions (2)

1. Beginning the Journey: Top Management Commitment (2)
2. Setting the Scope (3)
3. Pre-assessment (3)

4. Corrective Action from the Pre-assessment (2)

5. The Certification Audit (5)

6. TL 9000 Metrics (1)

7. Estimate your Client's Effort (2)

8. Review of the Approach (6)

## Some Facts About the Case Study Survey Respondents

The following information summarizes the characteristics of the case study respondents.

- Organizational size ranged from 35 in one organization to more than 12,350 employees at 15 separate locations in another organization.

- All organizationswere registered previously to ANSI/ISO/ASQC Q9001–1994 (8) or ANSI/ISO/ASQC Q9002–1994 (1).

- TL 9000 requirements already covered by the QMS ranged from 40 percent to more than 90 percent. $CSQP^{SM\ 5,6}$ (Customer Supplier Quality Program—$CSQP^{SM}$ is a service mark of Telcordia) and TickIT registrations were cited as factors by those at or above 85 percent.

- ISO 9001 and ISO 9002 registrars were used by 90 percent, but one company switched from its ISO 9001 registrar because the former registrar was not qualified for TL 9000 registration.

- Each organization obtained registrations in one of four TL 9000 categories—hardware (HW), software (SW), services (SV), and combined hardware/software (HW and SW).

- Sixty percent of TL 9000 registration assessments were done as part of a regularly scheduled ISO 9001 and ISO 9002 surveillance audit.

- Fifty percent relied on "in-house" internal auditors, with the others relying on internal auditors from other company facilities, a combination of internal and external auditors, or external auditors .

- Fifty percent had a pre-assessment done by the registrar.

- Four registrations covered only one facility, while one registration covered more than 12 facilities in four countries.

- Fifty percent plan to extend the scope of their TL 9000 registrations, and 10 percent more may do so.

- Thirty percent integrated TL 9000 with QS-9000- and/or ISO 14001-based management systems.

- Top management supported all registration efforts, with the source of approval ranging from the President/CEO to the VP/Director of Quality. The registrars agreed that from their perspective, top management support was strong.

As with ISO 9001 and ISO 9002 and other registration efforts, top management support and involvement is critical to TL 9000 implementation and registration. Two examples illustrate this:

1. In one organization, there was apprehension over the possibility of a large resource requirement, which raised questions about the importance of the program—particularly since management did not have customers requesting TL 9000 conformance. Once the benefits were explained, top management was very supportive of the effort.

2. In another organization, corporate-level management gave approval for TL 9000, but the implementation process ran into resistance from middle management in the facility. This required corporate management to strongly state the company's commitment to TL 9000 before the implementation team had complete internal support.

## OBTAINING MANAGEMENT COMMITMENT

Questions to address and discuss:

- What are the business benefits your organization will achieve by implementing TL 9000?

- Are customers requesting TL 9000 registration?

- Do major customers even know what TL 9000 is and what its benefits are?

- What are the estimated total resource impact and implementation costs?

- Are there any existing companywide goals that might conflict with or compete for resources with TL 9000 implementation and registration? Can TL 9000 leverage off any of those goals (for example, implementing a new product lifecycle tool)?

- Can TL 9000 be integrated with other initiatives, such as QS-9000, TickIT, or the Carnegie-Mellon Software Engineering Institute Capability Maturity Model?

A risk in having a lack of internal support within some areas of your company is that, when those who are initially resistant to the effort see the results and benefits, their attempts to catch up with the implementation effort could require additional resources. It's important to get the internal support at the start.

## Cost Vs. Value

Registration is often required by a customer, market and/or industry, but cost and value have been an issue for companies and other organizations deciding whether to implement management systems and seek registration to ISO 9000, QS-9000, ISO 14001 and/or other standards or requirements. Several survey questions asked about the cost and value of TL 9000 registration, and the responses provide useful perspectives.

Costs relating to implementation and registration take several forms, most notably the cost of hiring registrars, consultants, and trainers; the cost of committing internal resources to the implementation and maintenance efforts; and the cost of interruptions to a facility's operations during development, implementation, and registration. The respondents provided information concerning several key cost factors that need to be considered by companies preparing for or considering TL 9000 implementation and registration:

- *Registrar costs*—Sixty percent of respondents had their TL 9000 registration assessment as part of an ISO 9001 or ISO 9002 surveillance audit, thereby reducing registrar travel and other expenses. However, the audits were more costly because TL 9000 requirements and metrics required 2–7 extra auditor days, with most requiring 4–5 added days.

- *Implementation costs*—The approach to implementation varied. The number of full-time employees assigned ranged from zero to four or more with a median of two to three. Meanwhile, the size of the implementation team ranged from 3 to 41 with a median of 10–12. In the survey results, you can clearly see different approaches, from companies relying solely on a small core of full-time implementers to those relying on a team effort with no one committed to the effort full-time. Nevertheless, each approach requires the organization to remove employees from day-to-day operations on at least a part-time basis, which has a short-term affect on productivity.

- *External support*—Although two respondents relied, at least in part, on external support for internal auditing, none of the respondents reported using outside consultants to help with the implementation effort. Clearly, the respondents did not need consultant assistance because their participation in the pilot program and their membership in the QuEST Forum gave them access to industry resources. In addition, respondents had an inside understanding of TL 9000 since all companies participated in the drafting of the TL 9000 handbooks. Contrast this with most companies using ISO 9001 and/or ISO 9002 without actively participating in the revision of the ISO 9000 standards. As a result, many of these companies will rely on consultants to help them update their QMSs to ANSI/ISO/ASQ Q9001–2000. The need for consultants to help with TL 9000 implementation and registration is likely to remain limited for QuEST Forum members as long as the Forum continues to provide support for its members and involves them in TL 9000 development and revision efforts. Examples of QuEST Forum support are the case studies planned for publication in *The Informed Outlook*, including studies of Nortel Networks[3] and Tellabs Operations[4] that appeared in the May and August 2000 issues. Several more are expected to be published during the first half of 2001 and the QuEST Forum plans a series of best practice conferences.

- *Training*—Although critical to the successful implementation and use of a QMS, training does involve removing employees from day-to-day operations for at least a brief period of time. How much time was involved? Responses ranged from 12–30 minutes to one week for each full-time employee. Most respondents indicated that significant time was devoted to training the implementation team and other key QMS staff (for example, internal auditors), with everyone else receiving limited training. The answers seemed closely connected to whether the company had used Telcordia (formerly Bellcore) quality requirements and/or had been involved in the CSQP$^{SM}$ process. A company assessing how much training will be required needs to ask itself: "How well acquainted are our employees with the requirements similar to those of TL 9000?"

- *Total effort to obtain registration*—Pilot program participants were asked to estimate how much total staff time was required to achieve registration. Because of the newness of TL 9000 and its metrics requirements, only seven in nine provided responses, and their estimates ranged from two staff months to 4.3 staff years, with a median of one staff year. However, the estimates provided do not indicate how much of the effort had been devoted to maintaining the existing QMS and performance measurements systems.

There are a number of values that respondents attributed to TL 9000 implementation and registration. Some translate into direct cost savings and profitability improvements, while others translate into improved management of quality and the organization. The following lists identify those values of TL 9000 implementation and registration experienced by pilot organizations that have direct cost savings and resulted in improved management of quality within the organization.

---

### Direct Cost Savings/Profitability Benefits

- Per one organization, replacing CSQP$^{SM}$ with TL 9000 saves approximately $20,000/year.

- Collecting TL 9000 metrics saves money by eliminating multiple data collection.

- Compliance to TL 9000 enhances customer satisfaction and loyalty and differentiates the organization from its competition. This results in maintaining or improving market share.

- A reduction in registration and internal auditing costs when some organizations combined multiple ISO 9001/2 registrations into a single TL 9000 registration.

- Using metrics and benchmarking significantly improves products.

- An improvement in the new product introduction process.

- Leveraged past quality investments (ISO 9000, CSQP,$^{SM}$ RQMS, etc.).

> **Improved Management of Quality and Business Operations**
>
> - More consistent product lifecycle management:
>   - Better view of customer needs;
>   - Stronger emphasis on design control;
>   - Use of a best-practice approach;
>   - An improved relationship with key internal and external suppliers;
>   - An improvement in the product traceability processes;
>   - The use of quality metrics in benchmarking and improvement;
>   - The use of self-directed teams was enhanced.
> - Improved business management processes
>   - Implementation resulted in improvement of existing processes;
>   - Measured results (that is, metrics) provided decision making information to management;
>   - Registration enhanced internal confidence of the staff;
>   - Quality objectives previously believed unachievable were addressed;
>   - Enhanced the discipline required to document the steps taken to improve products;
>   - Enhanced the ability to replicate successes throughout the organization.

One respondent made clear that a key value of TL 9000 was in the metrics: "Our company has been collecting the metrics mandated by TL 9000 in some form or another for years; metrics are an important part of any effective quality system and are used by executives to make business decisions. [Collecting] the TL 9000 metrics will [result in] this review continuing to occur; [in addition], the metrics will now be included in industry benchmarks that will [result in] refined business decisions."

## Managing the Implementation Process

While there was variation in how case study survey respondents handled TL 9000 implementation and registration, the following "program" sums up common activities:

- Obtain commitment and support from upper management.
- Identify an executive sponsor for TL 9000 implementation.
- Select a registrar.
- Initially train one or two individuals in the contents of TL 9000. These individuals will then lead the gap analysis described below.
- Obtain input regarding the current status of the QMS.
- Set the scope of the TL 9000 certification. Make sure you understand the definitions of the product categories. Discuss your definition of scope with the registrar.
- Conduct an initial gap analysis of the existing QMS with respect to the TL 9000 requirements and metrics.
- Determine implementation and registration resource requirements and develop a project plan timeline.
- Discuss your implementation plans with your registrar.
- Organize the implementation team. As discussed previously, survey respondents took varying approaches to implementing TL 9000 which ranged from a small core group that did all the work to a large team with a part-time director. Factors to be considered by a company implementing TL 9000 are how well-established the ISO 9001/2-based QMS is, the resources required and complications faced previously when implementing ISO 9001/2, and the gaps in conformance to TL 9000 and the metrics. Depending on these factors, a company may choose to select a small group of full-time implementers or rely on a broad-based team of "part-time" implementers.
- Use training classes to bring the implementation team members and internal auditors up to speed. The purpose of implementation team training is to ensure that the team is consistently

implementing the TL 9000 "adders" and the new metrics throughout the organization. Because they have experience with ISO 9000, internal auditors will play a critical early role in gap analysis of the existing QMS against TL 9000's additional requirements and metrics. However, some requirements and metrics may be entirely new to them and the QMS.

- Conduct a detailed gap analysis of the ISO 9001/2-based QMS against all requirements and metrics of TL 9000. The implementation team needs to take the lead in reviewing existing documentation and metrics against TL 9000, with the internal auditors checking the QMS documentation as revisions are made.

- Develop a gap closure plan and ensure that all gaps are closed.

- Prepare the organization for the TL 9000 implementation and registration process. This involves communication with and training of all employees who affect quality so that they understand the changes to the QMS and support the use of TL 9000. Four steps were identified that are recommended for all companies:

    1. Develop a communications plan to introduce TL 9000 and explain implementation;

    2. Engage in communication activities, including presentations to senior management about TL 9000;

    3. Develop internal training and awareness materials appropriate to different staff levels; and

    4. Conduct in-house training of all management and associates.

- Inform customers of TL 9000 implementation and registration plans.

- Revise the QMS and conduct internal audits of the TL 9000 "adders" and metrics.

## Benefits of the Gap Analysis and Internal Audits

Results of gap analysis and internal auditing in the pilot organizations indicated that there are several problem areas to be addressed before achieving TL 9000 conformance. These include:

- Training was not adequate on revised and new processes;

- A lack of documentation (for example, metrics reporting system, disaster recovery, customer satisfaction, lifecycle model, and servicing);

- Establishment of an integrated metrics program, with alignment to RQMS[7] metrics where appropriate;
- Significant revision of quality manual;
- Creation of procedures related to software and training requirements;
- Development of processes to satisfy requirements for customer-supplier communication, collection of customer satisfaction data, and disaster recovery planning;
- Implementation of a lifecycle model (LCM);
- Development of a procedure for tracking design changes to specific products in the field.

## TL 9000 PROCESS DEVELOPMENT AND DOCUMENTATION

Depending on a company's existing QMS—is it registered to CSQP[SM] or does it simply conform to ISO 9001?—the number of new procedures that must be developed and documented will vary. However, among the survey respondents, the following processes most frequently were missing or in need of significant updating. The processes causing the most difficulty are listed first:

- TL 9000 metrics and data collection, verification, and reporting (Book Two, Sections 3.1 and 3.4.3)
- Documented quality improvement program (Book One, Section 4.21.1.C.1)
- Disaster recovery (Book One, Section 4.2.3.C.4)
- Advanced quality training (Book One, Section 4.18.C.5)
- Customer satisfaction reporting (Book One, Section 4.21.3.C.1)
- Customer-supplier communication (Book One, Section 4.21.2.C.2)
- Lifecycle model (Book One, Section 4.2.2.C.1)
- Tracking design changes (Book One, Section 4.4.9.H.1)
- Risk assessment (Book One, Section 4.4.2.C.1k [Project Plan] and 4.6.1.C.1b [Purchasing])

- Developing internal training (Book One, Section 4.18.C.1)
- Servicing adders (Book One, Section 4.19)
- Support software and tools management (Book One, Section 4.2.2.S.1)
- Computer resources for the target computer (Book One, Section 4.4.2.S.2)
- Integration planning (Book One, Section 4.4.2.S.3)

The amount and type of additional procedures required may dictate the makeup of the implementation team. In the pilot program, TL 9000 implementation teams membership varied with the size of the company and the status of its QMS conformance. While a few respondents indicated that the team represented a cross-section of the company, other teams only included one or more of the following types of employees: ISO 9000 internal auditors/coordinators; metrics SMEs (subject matter experts); quality system/quality assurance/software quality personnel; manufacturing/reliability engineers; production supervisors/operators. One respondent gave each team member a specific organizational role (for example, primary responsibility for training, auditing metrics, etc.), which helped ensure effective implementation. Four aspects of TL 9000 implementation deserve special attention. These are training, internal auditing, metrics, and the lifecycle model.

## Training

Training, both external and internal to companies implementing TL 9000, is critical. The QuEST Forum worked with Excel Partnership and STAT-A-MATRIX to develop sanctioned courses ranging from a general TL 9000 overview course to specific TL 9000 auditing and metrics courses. Ninety percent of the pilot organizations have sent employees to one or more of the courses. The other pilot organizations conducted their own internal training.

## Internal Auditing

Another critical process was internal auditing, which many survey respondents implemented according to the following general steps:

- Selection and training of the internal auditing team;
- Development of internal auditing schedules;

- Enhancement of existing (ISO 9001/2) internal audit checklists; and
- Use of internal audits for gap analysis.

## Metrics

Some metrics are common to all types of telecommunications companies, while others are specific to hardware, software, or service organizations. Metrics use and the collection, verifying, and reporting of data have already been noted as a difficulty for some respondents. This is largely due to the fact that in many cases the metrics required by TL 9000 did not match the metrics they had already been collecting (for example, they were using different data formulas and counting rules). Those respondents that did not have a problem with the metrics acknowledged that they have been using metrics very similar to those in TL 9000 (for example, RQMS) for years. Among those that had difficulty with metrics, the following are examples of what respondents identified as main obstacles:

- Obtaining data for each of the metrics from several different organizations within the company made the process of collecting data very complex. This complexity caused timing issues that could compromise data submission.

- Explaining the metrics to some managers and staff due to a lack of training was often very difficult.

- Developing a secure program to transmit data to UTD was difficult.

- Aligning an already well-established system with the new TL 9000 metrics and then getting everyone used to the new alignment and vocabulary was difficult.

## The Lifecycle Model

The ease or difficulty of implementing the lifecycle model (LCM) depended on the existing QMS. Respondents reported that a LCM was already in place in 40 percent of pilot facilities. Of the others, one, with the agreement of its registrar, did not implement the LCM because the scope only covered manufacturing and the organization had no control over design and development and use and replacement. Of those without an existing LCM, the development process was difficult, with the following problems encountered:

- One facility already had some LCM steps outlined, but "end-of-life" planning and migrations of certain systems created complications;

- One facility found the entire LCM process difficult and used a stage gate process;

- One facility adjusted its design control documentation but still had difficulty with the definition of LCM.

## PRE-ASSESSMENT AND REGISTRATION RESULTS

Half of the respondents underwent a pre-assessment audit. The benefits to those companies ranged from:

- Verification that the QMS was in good shape,

- To identifying areas of TL 9000 where the registrar differs from the company on interpretations,

- To pointing out areas where the company had overlooked something.

However, one respondent indicated that there were no benefits from the pre-assessment because the "registrar did not know enough about the standard."

What were the most significant findings indicated in the pre-assessments? Pre-assessments allowed organizations to work with their auditors to understand the interpretations of the adders. In some cases, because the auditors were as new to TL 9000 as the organization, there was a joint effort to formulate an understanding of the intent of the requirements. The following five major findings were reported by respondents:

1. Metrics reporting system was not documented;

2. Procedure for documenting disaster recovery was not established;

3. Procedure for communicating with selected customers was not established;

4. Inspection and test records were incomplete, not maintained, or not identified;

5. Training content for courses covering hazardous conditions had not yet been released.

When it came to registration results, three respondents stated they had zero nonconformances during the registration assessments. Only one reported a major nonconformance. Most of the nonconformances had already been identified as obstacles or difficult processes to comply with. The respondents resolved all significant nonconformances.

The registrars felt that the pilot program went well; there were no problems from their standpoint. One registrar stated that "this was as good and as smooth as any pilot program in which [they] have participated." Another stated that "both parties were determined to enable a first class effort."

# RECOMMENDATIONS TO ORGANIZATIONS IMPLEMENTING TL 9000

Eight key suggestions were mentioned:

1. Get senior management directly involved in the implementation process, not just lending their "support."

2. Get a better understanding of resource requirements.

3. Configure the implementation team to include a cross-section of the company.

4. Start the metrics program early.

5. Take more time. Some time allocation problems were the result of starting the pilot program while the TL 9000 handbooks were still being drafted. Most pilot organizations would have liked more time to make changes (update the ISO 9001-based QMS) and to take corrective action in response to the audits. Respondents indicated that they would have started training earlier; devoted more time to training, communication/awareness, gap analysis, and implementation; and allowed more time between the pre-assessment and registration assessment to take corrective actions.

6. Conduct a pre-assessment.

7. Increase customer awareness of the company's TL 9000 effort.

8. Take the opportunity to combine facilities under a single TL 9000 certificate.

Perhaps the best summary of the pilot program came from a registrar: "TL 9000 is straightforward and should pose no problem if organizations are committed. Organizations should start with the scope definition, product category table, and metrics gathering planning before diving into the registration process. Products and metrics drive the process and this must be well understood by the organization."

# END NOTES

1. *TL 9000 Quality System Requirements*, Book One, Release 2.5 (The QuEST Forum, 1999).
2. *TL 9000 Quality System Metrics*, Book Two, Release 2.5 (The QuEST Forum, 1999).
3. "For Nortel Networks, TL 9000 Means 30 Percent More." *The Informed Outlook* 5, no. 5: 1, 32–38 (May 2000).
4. Tellabs. *The Informed Outlook* 5, no. 8: 6–11 (August 2000).
5. GR-1202-CORE, *Generic Requirements for Customer Sensitive Quality Infrastructure* (Morristown, NJ: Telcordia Technologies, Issue 1, October 1995).
6. SR-3535, *Bellcore CSQP$^{SM}$ Program* (Morristown, NJ: Telcordia Technologies, Issue 1, October 1995).
7. GR-929-CORE, *Reliability and Quality Measurements for Telecommunications Systems (RQMS)* (Morristown, NJ: Telcordia Technologies, Issue 4, December 1998).

# 7
# Some Final Thoughts

The QuEST Forum was created as an experiment in cooperation among competing suppliers and their customers. The vision was to create a set of standards that could be used to improve the industry. The Forum was innovative in a number of ways, notably the creation of a metrics process that provides measurements for continual improvement.

Quality is an abstract activity carried out by every individual in an organization working at their best. Increased productivity and high customer satisfaction require established processes, effective training, and continual improvement. These principles are at the heart of TL 9000. They make good business sense and are the initial step in the long journey toward business excellence.

Telecommunications organizations worldwide have started to hearken to these principles. The QuEST Forum has experienced a phenomenal growth in the two and one-half years of its existence. Starting in January 1998 with 44 members, it had grown to 151 by December 31, 2000. And, as of December 31, 2000 there were 46 TL 9000 registrations, including organizations from Europe, Japan, Taiwan, China, and Hong Kong.

The Forum is now in the process of expanding its activities by working on the following new initiatives:

- Aligning the Registration Handbook with ANSI/ISO/ASQ Q9001–2000;

- Fine tuning the current measurements and evaluating the addition of other measures to the Metrics Handbook;

- Revising the training materials to align with change, especially due to alignment of the Registration Handbook with ANSI/ISO/ASQ Q9001–2000;

- Expanding global membership by holding meetings in Europe, Japan, Singapore, and Brazil;

- Expanding the industry membership to include the cable industry, the new service providers, and the new suppliers;

- Extending TL 9000 throughout supply chain;

- Developing the Business Excellence Acceleration Model (BEAM) aimed at continual improvement of processes, products, and services. A work group has been formed to develop BEAM based on business excellence models and guidance for excellence within the telecommunications industry; and

- Organizing an annual QuEST Forum Best Practices Symposium and developing a web site to communicate excellence within the industry.

In this book we have attempted to shed light on the contents of TL 9000 and have shared valuable tips on how to meet many of the requirements. Building a competitive advantage should be based on solid business strategy and a practical and efficient system of processes that yields desired measurable results. The QuEST Forum hopes to provide the nucleus of such a system.

# Appendix 1

## Tables of Adders to ANSI/ISO/ASQC Q9001–1994 and ANSI/ISO/ASQ Q9001–2000

Table A1.1 depicts the original source for each TL 9000 requirement. Requirements that are being introduced by the QuEST Forum and do not come from one of the sources are shown in the "QuEST Forum" column. The table is organized by the ANSI/ISO/ASQC Q9001–1994 requirements and contains the labels for the corresponding adders that appeared in the year 2000 edition of the document. Table A1.2 is similar except that it is organized by the ANSI/ISO/ASQ Q9001–2000 requirements and shows the corresponding TL 9000 adders based on the ANSI/ISO/ASQC Q9001–1994 requirements.

**Table A1.1** TL 9000 adders organized by ANSI/ISO/ASQC Q9001–1994 requirements.

| TL 9000 Book One | | Source Document | | | | | | |
|---|---|---|---|---|---|---|---|---|
| Adders to ANSI/ISO/ASQC Q9001–1994 | Adders to ANSI/ISO/ASQ Q9001–2000 | GR-1202 | GR-1252 | TR-179 | ISO 12207 | QuEST Forum | ISO 9000-3 | ISO 9004-2 |
| 4.1.1.C.1 | 5.4.1.C.1 | | | | | X | | |
| C-Note A | 5.4.2.C.1-Note 1 | | | | | | | |
| 4.2.2.C.1 | 7.1.C.1 | | | | 5.2.4.2 | | | |
| 4.2.2.S.1 | 7.1.S.3 | | | (R)-4.6-7 | | | | |
| 4.2.3.C.1 | 5.4.2.C.2 | R2-18 | | | | | | |
| 4.2.3.C.2 | 5.4.2.C.1 | R2-19 | | | | | | |
| 4.2.3.C.3 | 5.4.2.C.3 | R2-20 | | | | | | |

| TL 9000 Book One | | Source Document | | | | | | |
|---|---|---|---|---|---|---|---|---|
| Adders to ANSI/ISO/ASQC Q9001–1994 | Adders to ANSI/ISO/ASQ Q9001–2000 | GR-1202 | GR-1252 | TR-179 | ISO 12207 | QuEST Forum | ISO 9000-3 | ISO 9004-2 |
| 4.2.3.C.4 | 7.1.C.3 | | | | 6.2.6.1 | | | |
| 4.3.2 C-NOTE B | 7.2.2 C-NOTE 1 | | | | | | 4.3.2 | |
| 4.3.2 C-NOTE C | 7.2.2 C-NOTE 2 | | | | 5.1.5.1 | | | |
| 4.4.1 V-NOTE D | DELETED | | | | | | | |
| 4.4.1.C.1 | 7.3.1.C.2 | | | 4.1.3-4 | | | | |
| 4.4.2.C.1 | 7.3.1.C.1 | | | | 5.2.4.5 | | | |
| 4.4.2 C-NOTE E | 7.3.1.C.1-NOTE 1 | | | | | | | |
| 4.4.2 C-NOTE F | 7.3.1.C.1-NOTE 2 | | | | | | | |
| 4.4.2.C.2 | 7.3.1.C.3 | | | 3.7.1 | | | | |
| 4.4.2.C.2 C-NOTE G | 7.3.1.C.3-NOTE 1 | | | | | | | |
| 4.4.2.C.3 | 7.1.C.4 | | | | 5.5.6.1 | | | |
| 4.4.2.S.1 | 7.1.S.1 & 7.1.S.1-NOTE 1 | | | (R) 3.4.3-1 | | | | |
| 4.4.2.S.2 | 7.1.S.2 | | | (R) 3.4.3-3 | | | | |
| 4.4.2.S.3 | 7.3.1.S.1 | | | | 5.3.8.1 | | | |
| NEW | 7.3.1.S.1-NOTE 1 | | | | | | | |
| 4.4.2.S.4 | 7.3.1.S.2 | | | | 5.5.5.2-3 | | | |
| 4.4.3 C-NOTE H | 7.3.1 C.2-NOTE 1 | | | | | | | |
| 4.4.4.C.1 | 7.3.2.C.1 | R2-10 | | | | | | |
| 4.4.4.C.2 | 7.3.2.C.2 | | | | 5.3.2.1-2 | | | |
| 4.4.4.H.1 | 7.3.2.H.1 | | R2-12 | | | | | |
| 4.4.4.S.1 | 7.3.2.S.1 | | | | 5.3.4.1 | | | |
| 4.4.4.S.2 | 7.3.2.S.2 | | | | | | 4.4.4.c | |
| 4.4.5.S.1 | 7.3.3.S.1 | | | | | | 4.4.5 | |
| 4.4.5.V.1 | 7.3.3.V.1 | | | | | | | 6.2.3 |
| 4.4.7 C-NOTE I | DELETED | | | | | | | |
| 4.4.8 HV-NOTE J | DELETED | | | | | | | |
| 4.4.8. C - NOTE K | 7.3.6 C-NOTE 1 | | | | | | | |
| 4.4.8.H.1 | 8.2.4.H.1 | | R2-13 | | | | | |

## Table of Adders

| TL 9000 Book One | | Source Document | | | | | | |
|---|---|---|---|---|---|---|---|---|
| Adders to ANSI/ISO/ASQC Q9001–1994 | Adders to ANSI/ISO/ASQ Q9001–2000 | GR-1202 | GR-1252 | TR-179 | ISO 12207 | QuEST Forum | ISO 9000-3 | ISO 9004-2 |
| 4.4.8.H.2 | 8.2.4.H.2 | | R2-14 | | | | | |
| 4.4.8 H-Note L | 8.2.4.H.2-Note 1 | | | | | | | |
| 4.4.8.H.3 | 8.2.4.H.3 | | R2-15 | | | | | |
| 4.4.9.C.1 | 7.3.7.C.1 | | R2-8 | | | | | |
| 4.4.9.C.2 | 7.3.7.C.2 | | R2-11 | | | | | |
| 4.4.9.H.1 | DELETED | | R2-16 | | | | | |
| 4.4.9.H.2 | 7.3.7.H.1 | | | | | X | | |
| 4.4.9 H-NOTE M | Included in 7.3.7.H.1 | | | | | | | |
| 4.4.9.V.1 | 7.5.1.V.2 | | | | | X | | |
| 4.5.1.S.1 | 4.2.3.C.1 | | | | | X | | |
| 4.6.1.C.1 | 7.4.1.C.1 | | R2-20 | | 5.1.1.8; 5.1.3.1 | | | |
| 4.6.1 C-NOTE O | 7.4.1.C.1-NOTE 1 | | | | | | | |
| 4.8.H.1 | 7.5.3.H.1 | | | | | X | | |
| 4.8.H.2 | 7.5.3.H.2 | | | | | X | | |
| 4.8.HS.1 | 7.1.HS.1 | | | | 6.2.1.1–6.2.2.1 | | | |
| 4.8.HS.1 HS NOTE P: | 7.1.HS.1-NOTE 1: | | | | | | | |
| 4.8.HS.2 | 7.5.3.HS.1 | | | | | | 4.8 | |
| 4.9.H.1 | DELETED | | R2-25 | | | | | |
| 4.9.HV.1 | 7.5.2.HV.1 | | R2-26 | | | | | |
| 4.9.HV.2 | 6.2.2.HV.1 | R2-37 | | | | | | |
| 4.9.HV.3 | DELETED | R2-38 | | | | | | |
| 4.9.S.1 | 7.5.1.S.3 | | | | | | 4.9 | |
| 4.9.S.2 | 7.3.6.S.1 | | | (R) 3.9.1-2,3 | 5.5.5.3, 5.5.6.2, 6.2.6.1 | | | |
| 4.9.V.1 | 7.5.1.V.1 | | | | | X | | |
| 4.9.V.2 | 7.1.V.1 | | | | 5.2.4.5 | | | |
| 4.10.1 C-NOTE Q | DELETED | | | | | | | |
| 4.10.1.HV.1 | 8.2.4.HV.1 | | R2-5 | | | | | |
| 4.10.1.S.1 | 8.2.4.S.1 | | | 3.7.1,2 | | | | |
| 4.10.4.H.1 | 8.2.4.H.4 | | R2-28 | | | | | |
| 4.10.4.H.2 | 7.5.5.HS.1 | | | | | X | | |
| 4.10.4.H.2 H-NOTE R | 7.5.5. HS.1-NOTE 1 | | | | | | | |
| 4.10.5.HV.1 | 8.2.4.HV.2 | | R2-42 | | | | | |

| TL 9000 Book One | | Source Document | | | | | | |
|---|---|---|---|---|---|---|---|---|
| Adders to ANSI/ISO/ASQC Q9001–1994 | Adders to ANSI/ISO/ASQ Q9001–2000 | GR-1202 | GR-1252 | TR-179 | ISO 12207 | QuEST Forum | ISO 9000-3 | ISO 9004-2 |
| 4.11.2.H.1 | 7.6.H.1 | | R2-31 | | | | | |
| 4.13.2.C.1 | 8.4.C.1 | | | | | X | | |
| 4.14.1 C-NOTE S | 8.5.1 C-NOTE 1 | | | | | | | |
| 4.14.2 C-Note T | 8.5.2. C-Note 1 | | | | | | | |
| 4.14.2 C-Note U | 8.5.2. C-Note 2 | | | | | | | |
| 4.15.1.C.1 | 6.4.C.1 | | R2-38 | | | | | |
| 4.15.1.C.2 | 7.5.5.C.1. | | R2-39 | | | | | |
| 4.15.2.S.1 | 7.5.5.S.1 | | | | | | 4.15.2; 4.15.6 | |
| 4.15.3.H.1 | 7.5.5.H.1 | | R2-40 | | | | | |
| 4.15.6.S.1 | 7.5.1.S.2 | | | 3.9.4-2 | | | | |
| 4.18.C.1 | 6.2.2.C.1 | R2-36 | | | | | | |
| 4.18.C.2 | 6.2.2.C.2 | R2-34 | | | | | | |
| 4.18.C.3 | 6.2.2.C.3 | R2-32, 35 | | | | | | |
| 4.18.C.4 | 6.2.2.C.4 | | | | | X | | |
| 4.18.C.5 | 6.2.2.C.5 | | R2-56 | | | | | |
| 4.18.C.6 | 6.2.2.C.6 | | | | | X | | |
| 4.19.C.1 | 7.5.1.C.1 | | R2-52 | | | | | |
| 4.19.C.2 | 7.5.1.C.2 | R2-16 | | | | | | |
| 4.19.C.3 | 7.2.3.C.1 | | R2-54 | | | | | |
| 4.19.C.4 | 7.2.3.C.2 & 7.2.3.C.2-NOTE 1 | | | 4.10.2-7,8 | | | | |
| 4.19.C.5 | 7.2.3.C.3 | | | 4.10.2-6 | | | | |
| NEW | 7.2.3.C.4 | | | | | X | | |
| 4.19.H.1 | 7.2.3.H.1 | | | | | X | | |
| 4.19.HS.1 | 7.5.1.HS.1 | R2-17 | | | | | | |
| 4.19.HS.2 | 7.3.7.HS.1 | | | 4.10.2-9 | | | | |
| 4.19.HS.3 | 7.5.1.HS.2 | | | | 5.3.12, 5.3.12.2, 5.5.5.4-5, 5.5.6.3-5 | | | |
| 4.19.S.1 | 7.5.1.S.1 | | | 3.10.3-1 | | | | |
| 4.19.S.2 | 8.5.2.S.1 | | | 4.10.2.5 | | | | |
| 4.20.1.C.1 | 8.2.3.C.1 | | | | | | 4.20 | |
| 4.21.1.C.1 | 8.5.1.C.1 | | R2-57 | | | | | |
| 4.21.1.C.1 C -NOTE V | 8.5.1 C-NOTE 2 | | | | | | | |

## Table of Adders

| TL 9000 Book One | | Source Document | | | | | | |
|---|---|---|---|---|---|---|---|---|
| Adders to ANSI/ISO/ASQC Q9001–1994 | Adders to ANSI/ISO/ASQ Q9001–2000 | GR-1202 | GR-1252 | TR-179 | ISO 12207 | QuEST Forum | ISO 9000-3 | ISO 9004-2 |
| 4.21.1.C.2 | 8.5.1.C.2 | R2-26 | | | | | | |
| 4.21.1.C.3 | 5.5.3.C.1 | R2-31 | | | | | | |
| 4.21.2.C.1 | 5.2.C.1 | R2-7 | | | | | | |
| 4.21.2.C.2 | 5.2.C.2 | R2-9, 15,18 | | | | | | |
| 4.21.2 C-Note W | 5.2.C.2-Note 1 | | | | | | | |
| 4.21.3.C.1 | 8.2.1.C.1 | R2-46 | | | | | | |
| 4.21.3.H.1 | 8.4.H.1 | | R2-46 | | | | | |
| 4.21.3.V.1 | 8.4.V.1 | | R2-46 | | | | | |
| 4.21.4.C.1 | 7.1.C.2 | | | | | X | | |
| 4.21.4 C-NOTE X | 7.1.C.2-NOTE 1 | | | | | | | |

**Table A1.2** TL 9000 adders organized by ANSI/ISO/ASQ Q9001–2000 requirements

| Adders to ANSI/ISO/ASQ Q9001–2000 | Adders to ANSI/ISO/ASQC Q9001–1994 | GR-1202 | GR-1252 | TR-179 | ISO 12207 | QuEST Forum | ISO 9000-3 | ISO 9004-2 |
|---|---|---|---|---|---|---|---|---|
| 4.2.3.C.1 | 4.5.1.S.1 | | | | | X | | |
| 5.2.C.1 | 4.21.2.C.1 | R2-7 | | | | | | |
| 5.2.C.2 | 4.21.2.C.2 | R2-9, 15,18 | | | | | | |
| 5.2.C.2-Note 1 | 4.21.2 C-Note W | | | | | | | |
| 5.4.1.C.1 | 4.1.1.C.1 | | | | | X | | |
| 5.4.2.C.1 | 4.2.3.C.2 | R2-19 | | | | | | |
| 5.4.2.C.1-Note 1 | 4.1.1.C-Note A | | | | | | | |
| 5.4.2.C.2 | 4.2.3.C.1 | R2-18 | | | | | | |
| 5.4.2.C.3 | 4.2.3.C.3 | R2-20 | | | | | | |
| 5.5.3.C.1 | 4.21.1.C.3 | R2-31 | | | | | | |
| DELETED | 4.9.HV.3 | R2-38 | | | | | | |
| 6.2.2.C.1 | 4.18.C.1 | R2-36 | | | | | | |
| 6.2.2.C.2 | 4.18.C.2 | R2-34 | | | | | | |
| 6.2.2.C.3 | 4.18.C.3 | R2-32, 35 | | | | | | |
| 6.2.2.C.4 | 4.18.C.4 | | | | | X | | |
| 6.2.2.C.5 | 4.18.C.5 | | R2-56 | | | | | |
| 6.2.2.C.6 | 4.18.C.6 | | | | | X | | |

## Appendix 1

| Adders to ANSI/ISO/ASQ Q9001–2000 | Adders to ANSI/ISO/ASQC Q9001–1994 | GR-1202 | GR-1252 | TR-179 | ISO 12207 | QuEST Forum | ISO 9000-3 | ISO 9004-2 |
|---|---|---|---|---|---|---|---|---|
| 6.2.2.HV.1 | 4.9.HV.2 | R2-37 | | | | | | |
| 6.4.C.1 | 4.15.1.C.1 | | R2-38 | | | | | |
| 7.1.C.1 | 4.2.2.C.1 | | | | 5.2.4.2 | | | |
| 7.1.C.2 | 4.21.4.C.1 | | | | | X | | |
| 7.1.C.2-NOTE 1 | 4.21.4 C-NOTE X | | | | | | | |
| 7.1.C.3 | 4.2.3.C.4 | | | | 6.2.6.1 | | | |
| 7.1.C.4 | 4.4.2.C.3 | | | | 5.5.6.1 | | | |
| 7.1.HS.1 | 4.8.HS.1 | | | | 6.2.1.1–6.2.2.1 | | | |
| 7.1.HS.1-NOTE 1: | 4.8.HS.1 HS NOTE P: | | | | | | | |
| 7.1.S.1 & 7.1.S.1-NOTE 1 | 4.4.2.S.1 | | | (R) 3.4.3-1 | | | | |
| 7.1.S.2 | 4.4.2.S.2 | | | (R) 3.4.3-3 | | | | |
| 7.1.S.3 | 4.2.2.S.1 | | | (R)-4.6-7 | | | | |
| 7.1.V.1 | 4.9.V.2 | | | | 5.2.4.5 | | | |
| 7.2.2 C-NOTE 1 | 4.3.2 C-NOTE B | | | | | | 4.3.2 | |
| 7.2.2 C-NOTE 2 | 4.3.2 C-NOTE C | | | | 5.1.5.1 | | | |
| 7.2.3.C.1 | 4.19.C.3 | | R2-54 | | | | | |
| 7.2.3.C.2 & 7.2.3.C.2-NOTE 1 | 4.19.C.4 | | | 4.10.2-7,8 | | | | |
| 7.2.3.C.3 | 4.19.C.5 | | | 4.10.2-6 | | | | |
| 7.2.3.C.4 | NEW | | | | | X | | |
| 7.2.3.H.1 | 4.19.H.1 | | | | | X | | |
| 7.3.1.C.1 | 4.4.2.C.1 | | | | 5.2.4.5 | | | |
| 7.3.1 C.1-NOTE 1 | 4.4.2 C-NOTE E | | | | | | | |
| 7.3.1.C.1-NOTE 2 | 4.4.2 C-NOTE F | | | | | | | |
| 7.3.1.C.2 | 4.4.1.C.1 | | | 4.1.3-4 | | | | |
| 7.3.1.C.2-NOTE 1 | 4.4.3 C-NOTE H | | | | | | | |
| 7.3.1.C.3 | 4.4.2.C.2 | | | 3.7.1 | | | | |
| 7.3.1.C.3-NOTE 1 | 4.4.2.C.2 C-NOTE G | | | | | | | |
| 7.3.1.S.1 | 4.4.2.S.3 | | | | 5.3.8.1 | | | |
| 7.3.1.S.1-NOTE 1 | NEW | | | | | | | |

| Adders to ANSI/ISO/ASQ Q9001–2000 | Adders to ANSI/ISO/ASQC Q9001–1994 | GR-1202 | GR-1252 | TR-179 | ISO 12207 | QuEST Forum | ISO 9000-3 | ISO 9004-2 |
|---|---|---|---|---|---|---|---|---|
| 7.3.1.S.2 | 4.4.2.S.4 | | | | 5.5.5.2-3 | | | |
| 7.3.2.C.1 | 4.4.4.C.1 | R2-10 | | | | | | |
| 7.3.2.C.2 | 4.4.4.C.2 | | | | 5.3.2.1-2 | | | |
| 7.3.2.H.1 | 4.4.4.H.1 | | R2-12 | | | | | |
| 7.3.2.S.1 | 4.4.4.S.1 | | | | 5.3.4.1 | | | |
| 7.3.2.S.2 | 4.4.4.S.2 | | | | | | 4.4.4.c | |
| 7.3.3.S.1 | 4.4.5.S.1 | | | | | | 4.4.5 | |
| 7.3.3.V.1 | 4.4.5.V.1 | | | | | | | 6.2.3 |
| 7.3.6 C-NOTE 1 | 4.4.8. C - NOTE K | | | | | | | |
| 7.3.6.S.1 | 4.9.S.2 | | | (R) 3.9.1-2,3 | 5.5.5.3, 5.5.6.2, 6.2.6.1 | | | |
| 7.3.7.C.1 | 4.4.9.C.1 | | R2-8 | | | | | |
| 7.3.7.C.2 | 4.4.9.C.2 | | R2-11 | | | | | |
| 7.3.7.HS.1 | 4.19.HS.2 | | | 4.10.2-9 | | | | |
| 7.3.7.H.1 | 4.4.9.H.2 & 4.4.9 H-NOTE M | | | | | X | | |
| 7.4.1.C.1 | 4.6.1.C.1 | | R2-20 | | 5.1.1.8; 5.1.3.1 | | | |
| 7.4.1.C.1-NOTE 1 | 4.6.1 C-NOTE O | | | | | | | |
| 7.5.1.C.1 | 4.19.C.1 | | R2-52 | | | | | |
| 7.5.1.C.2 | 4.19.C.2 | R2-16 | | | | | | |
| 7.5.1.HS.1 | 4.19.HS.1 | R2-17 | | | | | | |
| 7.5.1.HS.2 | 4.19.HS.3 | | | | 5.3.12, 5.3.12.2, 5.5.5.4-5, 5.5.6.3-5 | | | |
| 7.5.1.S.1 | 4.19.S.1 | | | 3.10.3-1 | | | | |
| 7.5.1.S.2 | 4.15.6.S.1 | | | 3.9.4-2 | | | | |
| 7.5.1.S.3 | 4.9.S.1 | | | | | | 4.9 | |
| 7.5.1.V.1 | 4.9.V.1 | | | | | X | | |
| 7.5.1.V.2 | 4.4.9.V.1 | | | | | X | | |
| 7.5.2.HV.1 | 4.9.HV.1 | | R2-26 | | | | | |
| 7.5.3.HS.1 | 4.8.HS.2 | | | | | | 4.8 | |
| 7.5.3.H.1 | 4.8.H.1 | | | | | X | | |
| 7.5.3.H.2 | 4.8.H.2 | | | | | X | | |
| 7.5.5.C.1. | 4.15.1.C.2 | | R2-39 | | | | | |
| 7.5.5.H.1 | 4.15.3.H.1 | | R2-40 | | | | | |

# Appendix 1

| Adders to ANSI/ISO/ASQ Q9001–2000 | Adders to ANSI/ISO/ASQC Q9001–1994 | GR-1202 | GR-1252 | TR-179 | ISO 12207 | QuEST Forum | ISO 9000-3 | ISO 9004-2 |
|---|---|---|---|---|---|---|---|---|
| 7.5.5.HS.1 | 4.10.4.H.2 | | | | | X | | |
| 7.5.5. HS.1-NOTE 1 | 4.10.4.H.2 H-NOTE R | | | | | | | |
| 7.5.5.S.1 | 4.15.2.S.1 | | | | | | 4.15.2; 4.15.6 | |
| 7.6.H.1 | 4.11.2.H.1 | | R2-31 | | | | | |
| 8.2.1.C.1 | 4.21.3.C.1 | R2-46 | | | | | | |
| 8.2.3.C.1 | 4.20.1.C.1 | | | | | | | 4.20 |
| 8.2.4.H.1 | 4.4.8.H.1 | | R2-13 | | | | | |
| 8.2.4.H.2 | 4.4.8.H.2 | | R2-14 | | | | | |
| 8.2.4.H.2-Note 1 | 4.4.8 H-Note L | | | | | | | |
| 8.2.4.H.3 | 4.4.8.H.3 | | R2-15 | | | | | |
| 8.2.4.H.4 | 4.10.4.H.1 | | R2-28 | | | | | |
| 8.2.4.HV.1 | 4.10.1.HV.1 | | R2-5 | | | | | |
| 8.2.4.HV.2 | 4.10.5.HV.1 | | R2-42 | | | | | |
| 8.2.4.S.1 | 4.10.1.S.1 | | | 3.7.1,2 | | | | |
| 8.4.C.1 | 4.13.2.C.1 | | | | | X | | |
| 8.4.H.1 | 4.21.3.H.1 | | R2-46 | | | | | |
| 8.4.V.1 | 4.21.3.V.1 | | R2-46 | | | | | |
| 8.5.1.C.1 | 4.21.1.C.1 | | R2-57 | | | | | |
| 8.5.1 C-NOTE 1 | 4.14.1 C-NOTE S | | | | | | | |
| 8.5.1 C-NOTE 2 | 4.21.1.C.1 C-NOTE V | | | | | | | |
| 8.5.1.C.2 | 4.21.1.C.2 | R2-26 | | | | | | |
| 8.5.2. C-Note 1 | 4.14.2 C-Note T | | | | | | | |
| 8.5.2. C-Note 2 | 4.14.2 C-Note U | | | | | | | |
| 8.5.2.S.1 | 4.19.S.2 | | | 4.10.2.5 | | | | |
| DELETED | 4.4.1 V-NOTE D | | | | | | | |
| DELETED | 4.4.7 C-NOTE I | | | | | | | |
| DELETED | 4.4.8 HV-NOTE J | | | | | | | |
| DELETED | 4.4.9.H.1 | | R2-16 | | | | | |
| DELETED | 4.9.H.1 | | R2-25 | | | | | |
| DELETED | 4.10.1 C-NOTE Q | | | | | | | |

# Appendix 2
# TL 9000 Registration and Accreditation

## INTRODUCTION

As you pursue TL 9000 registration, you'll find striking similarities between it and other certification efforts your company may have undertaken—particularly ISO 9000. The TL 9000 registration period extends over three years, during which time the entire system is examined. The scope of the registration may include hardware, software, services, or any combination of the three, and could apply to part or all of an organization. The scope you choose will be clearly identified on your registration certificate. The number of days required for the audit is based on established guidelines, and any nonconformances must be resolved before you receive your certificate. Trained auditors with expertise in the telecommunications industry will perform the assessment of your company, and registrars and accreditation bodies must follow criteria set forth by the QuEST Forum.

On a different note, the QuEST Forum recognizes that there are many well-established quality systems in place. Companies with mature quality systems may choose to pursue the alternative method for registration to TL 9000 that takes advantage of their internal audit systems.

## GETTING STARTED

To begin your journey, log in to the QuEST Forum web site at www.questforum.org. The TL 9000 Registration Process page describes how to obtain identification numbers and passwords, and allows you to download the data submission software package. The web site is a terrific source of information. Some useful links include a general overview of TL 9000, registration instructions and templates, a list of QuEST Forum member companies, and a list of TL 9000 qualified registrars.

Select your registrar early in the implementation process. Choose a registrar based on their ability to meet your business needs and to provide you with qualified auditors. Discuss your scope and approach to TL 9000 with your registrar to ensure that you and your registrar are in agreement—you must prevent surprises during the initial assessment.

Consider the following checklist for TL 9000 audit readiness:

- Registrar/auditors selected and engaged;
- TL 9000 requirements effectively implemented;
- Metrics data submitted and confirmation report received;
- At least one management review performed; and
- At least one internal audit cycle completed.

## The Initial Audit

The audit is intended to verify that the TL 9000 requirements have been effectively implemented. The auditors will be experienced ISO 9000 auditors, and all will have attended and successfully completed QuEST sanctioned training to ensure consistency of interpretation of the requirements. Additionally, at least one member of the audit team must have experience in the telecommunications industry.

The number of days required for the assessment will depend on the size of the facility and number of employees. An audit days table based on ISO Guide 62 is located on the QuEST Forum web site. The table lists the minimum number of on-site days required for both the registration and surveillance audits. If the number of auditor days proposed by your registrar is more than .5 days less than what is required by the audit days table, your registrar must submit a deviation request to the RAB within five days of providing you with a quote. The accreditation body will provide a written response within 10 days.

When your audit takes place, the auditors will identify opportunities for improvement along with nonconformities, and document them in a report that they will provide to you within 45 days of the audit. Any major or systemic nonconformances must be resolved prior to your certificate being issued. Your registrar is required to have a process in place to settle any disputes you may have regarding matters of interpretation and nonconformance.

If you have chosen to utilize the services of a consultant for your implementation efforts, be aware that although they may be present during the assessment, their role will be strictly limited to that of an observer.

## Migration to TL 9000

If your company maintains a current registration to an existing quality system standard, you may follow a migration path to achieve TL 9000 registration. The QuEST Forum currently recognizes ANSI/ISO/ASQC Q9001–1994, ANSI/ISO/ASQC Q9002–1994, ANSI/ISO/ASQ Q9001–2000 and QS-9000 quality system registrations. Other registrations such as TickIT will be recognized as the need arises. If your company is currently registered to QS-9000 for example, your registrar will skip all of the ISO 9001 and customer satisfaction requirements on its *initial* assessment of your quality management system. Subsequent assessments will incorporate all of the TL 9000 requirements. If however, your company is not currently registered to any of the above quality standards, *all* of the TL 9000 requirements will be audited during the initial assessment.

## Publicizing Your Registration

Once you have achieved TL 9000 registration, you may publicize that fact. Ensure that your advertising states that your quality system has been registered to TL 9000, and does not in any way suggest that your products have been registered. The TL 9000 certification mark may not be used on any product or packaging labeling. Further, you must take the scope of your registration into account when deciding how and where to advertise so you do not imply that nonregistered organizations within your company are included within the scope of registration. You may use the TL 9000 certification mark on stationery and promotional materials. Any misuse of the certification mark is grounds for withdrawal of your certification.

## Your Registrar

Your TL 9000-approved registrar is required to meet stringent criteria to ensure quality and consistency in the registration process. All approved registrars must follow the requirements set by the QuEST Forum on how to perform TL 9000 registrations, and must comply with the "Code of Practice for TL 9000 Registrars" which outlines ground rules for the registration process.

The approved registrars are required to have in-house telecommunications and TL 9000 expertise to better support your efforts. Each registrar must have at least one member of their governing board or council of experts with telecommunications experience who has successfully passed the QuEST sanctioned TL 9000 training.

As part of the registrar qualification process, a QuEST Forum-recognized accreditation body witnesses each registrar during one of its first six TL 9000 audits. Registrars may not use the TL 9000 notation on registration certificates prior to approval, after which the registrars will receive a certificate, or some other form of notification from the accreditation body. All recognized accredited registrars are listed on the QuEST Forum web site in the "TL 9000 Qualified Registrar Listing." Accreditation bodies are subject to similar qualification criteria and must meet the requirements established by the QuEST Forum.

## Alternate Method for Maintaining Certification

If you have a mature quality management system, you may opt for the alternative method for TL 9000 registration. The purpose of the alternative method is to eliminate audit duplication and to ensure that your third party audits are effective by allowing your registrar to utilize the results of your internal audit system as a basis for their assessments. This approach empowers your company by shifting the responsibility for verifying the effectiveness of your quality management system from the registrar to your company. The extent to which internal audits may be used as a substitute for external audits will be different for each organization and depends on various factors, including the maturity of the quality system and the ability of the registrar to verify internal audit results. Advantages to this approach include reduced cost due to fewer external auditor days on site and less disruption to the organization. Recognize that your registrar will still conduct on site audits, but on a limited basis by sampling both TL 9000 requirements and sites.

If you desire to pursue the alternative method, be aware of stringent qualification criteria. Your TL 9000 compliant quality system must have been in place for at least three years. You must operate an effective internal audit system that adheres to ISO 10011 and have a robust management review process in which goals and objectives are reviewed and updated, and corrective and preventive action is a routine part of the process. Your management must demonstrate their commitment to quality and to improvements of quality objectives and results over at least a two year period. You must also be able to demonstrate customer satisfaction.

Your registrar will base their audits on the effectiveness of your quality performance over the last three years by considering earlier surveillance audit results, results from independent customer satisfaction surveys and customer complaints. If your company fails to meet the alternative method criteria, your registrar will default to the regular TL 9000 registration process.

Your registrar and accreditation bodies must meet similarly challenging criteria. Registrars must demonstrate the maturity of their own business systems by designing a process to support the alternative method approach, and accreditation bodies by their ability to verify this capability. Registrars must be accredited to ISO/IEC Guide 62 for at least three years, and must be able to demonstrate improvement of service quality objectives and results over a period of three years.

## CONCLUSION

As you journey towards TL 9000 registration, keep yourself informed of the guidelines and resources on the QuEST Forum web site. Develop a relationship with your registrar early in the implementation process to build an effective partnership. Expect that your auditors, registrar, and accreditation bodies have met the criteria established by the QuEST Forum to ensure quality TL 9000 assessments. Follow the TL 9000 migration paths where appropriate to save time and money, and strive for meeting the qualification criteria for the alternative method approach to maintaining TL 9000 registration. Good luck on your TL 9000 endeavors.

# Bibliography

Alcatel Registers All Products, Services, Facilities to TL 9000. *Quality System Update* 10, no. 7:4, 17. July 2000.

Anderson, Stewart. TL 9000: Reshaping the Telecommunications Industry. *The Informed Outlook* 4, no. 8:10–12. August 1999.

ANSI/ISO/ASQC Q10011-1–1994. *Guidelines for Auditing Quality Systems—Auditing.* Milwaukee, WI, American Society for Quality Control. 1994.

ANSI/ISO/ASQC Q10011-2–1994. *Guidelines for Auditing Quality Systems—Qualification Criteria for Quality Systems Auditors.* Milwaukee, WI, American Society for Quality Control. 1994.

ANSI/ISO/ASQC Q10011-3–1994. *Guidelines for Auditing Quality Systems—Management of Audit Programs.* Milwaukee, WI, American Society for Quality Control. 1994.

ANSI/ISO/ASQC Q9004-2–1994. *Quality Management and Quality System Elements—Part 2: Guidelines for Services.* Milwaukee, WI, American Society for Quality Control. 1994.

ANSI/ISO/ASQ Q9000-3–1997. *Quality Management and Quality Assurance Standards—Part 3: Guidelines for the Application of ISO 9001:1994 to the Development, Supply, Installation, and Maintenance of Computer Software.* Second Edition. Milwaukee, WI, American Society for Quality. 1997.

Aycock, Galen, Jean-Normand Drouin, and Thomas Yohe. TL 9000 Performance Metrics to Drive Improvement. ASQ *Quality Progress* 32, no. 7:41–45. July 1999.

Caplan, Frank. *The Quality System.* Second Edition. Chilton Book Company. 1990.

Evans, Michael W. and John J. Marciniak. *Software Quality Assurance & Management.* New York: John Wiley & Sons. 1987.

For Nortel Networks, TL 9000 Means 30 Percent More. *The Informed Outlook* 5, no. 5:1, 32–38. May 2000.

Gildersleeve, James P. TL 9000 Quality System Requirements Rooted in ISO 9001. ASQ *Quality Progress* 32, no. 6:67–68. June 1999.

GR-1202-CORE, *Generic Requirements for Customer Sensitive Quality Infrastructure.* Morristown, NJ: Telecordia Technologies, Issue 1, October 1995.

Hutchison, Eugene and Sandford Liebesman. Telecom Industry Aims to Save Billions with ISO 9000-Based Standard. *ISO 9000 News* 7, no. 6:1, 20–24. November/December 1998.

Jarvis, Alka and Vern Crandall. *Inroads To Software Quality.* Englewood Cliffs, NJ: Prentice Hall. 1997.

Jarvis, Alka S. *Applying Software Quality.* The Seventh International Software Quality Week. San Francisco, CA. May 1994.

———. *Exploring the Needs of A Developer and A Tester.* Quality Conference. Santa Clara, CA. April 1995.

———. *Improvements Process Based on Metrics.* Third Annual Software Quality Conference. Milpitas, CA. May 1993.

Kehoe, Ray and Alka S. Jarvis. *ISO 9000-3—A Tool for Software Product and Process Improvement.* Springer Verlag, 1995.

Liebesman, Sandford. TL 9000—An Update on the Metrics Process. ASQ *Quality Progress* 33, no. 8:100–102. August 2000.

Malec, Henry. TL 9000 Database Repository and Metrics. *The Informed Outlook* 4, no. 6:4. June 1999.

Obolewicz, Steve. Getting to Know TL 9000. *Quality System Update* 9, no. 6:16–17. June 1999.

Participation in the TL 9000 Pilot Program: An Inside View. *The Informed Outlook* 4, no. 10:3–6. October 1999.

TL 9000 = ISO 9001 + 83 Telecom Sector Requirements + Metrics + Dedicated Data Base. *ISO 9000 News* 8, no. 5:28, 29, 31. September/October 1999.

TL 9000 Metrics: The Key to Continuous Improvement. *The Informed Outlook* 4, no. 5:24, 25. May 1999.

TL 9000 Pilot Program Case Studies Enhance Focus. *The Informed Outlook* 5, no. 4:1, 27–33. April 2000.

*TL 9000 Quality System Requirements.* Book One, Release 2.5. The QuEST Forum. 1999.

*TL 9000 Quality System Metrics.* Book Two, Release 2.5. The QuEST Forum. 1999.

Welch, Steve. 21 Voices for the 21$^{st}$ Century. ASQ *Quality Progress* 33, no. 1:31–32. January 2000.

# Index

Numbers in italics refer to illustrations.

## A

adders
    Advanced Quality Training (Adder 4.18.C.5), 41
    Anti-Static Protection (Adder 4.15.1.C.2), 38
    Change Management Process (Adder 4.4.9.C.1), 33
    Component Changes (Adder 4.4.9.H.2), 60
    Computer Resources (Adder 4.4.2.S.2), 66
    Configuration Management (CM) Plan (Adder 4.8.HS.1), 53
    Content of Requirements (Adder 4.4.4.H.1), 58
    Content of Testing (Adder 4.4.8.H.2), 59
    Control of Customer-Supplied Documents and Data (Adder 4.5.1.S.1), 69
    Course Development (Adder 4.18.C.1), 39
    Customer and Subcontractor Input (Adder 4.4.4.C.1), 31
    Customer Involvement (Adder 4.2.3.C.1), 22
    Customer Satisfaction (Adder 4.21.3.C.1), 48
    Customer-Supplier Communication (Adder 4.21.2.C.2), 47
    Design Output (Adder 4.4.5.S.1), 68
    Design Requirements (Adder 4.4.4.C.2), 31
    Deterioration (Adder 4.15.3.H.1), 64
    Disaster Recovery (Adder 4.2.3.C.4), 24
    Emergency Service (Adder 4.19.HS.1), 54
    Employee Participation (Adder 4.21.1.C.2), 46
    Employee Skills List (Adder 4.9.HV.3), 56
    End of Life Planning (Adder 4.4.2.C.3), 30
    ESD Training (Adder 4.18.C.4), 41
    Estimation (Adder 4.4.2.S.1), 66
    Field Performance Data (Adder 4.21.3.H.1), 64–65
    Frequency of Testing (Adder 4.4.8.H.3), 59
    Identification of Software Requirements (Adder 4.4.4.S.1), 68
    Identified Equipment (Adder 4.11.2.H.1), 63
    Informing Customers (Adder 4.4.9.C.2), 34
    Inspection and Test Documentation (Adder 4.10.1.HV.1), 57
    Inspection and Test Records (Adder 4.10.5.HV.1), 57
    Inspection and Testing (Adder 4.9.H.1), 62
    Installation Plan (Adder 4.19.HS.3), 55
    Integration Planning (Adder 4.4.2.S.3), 67
    Lifecycle Model (Adder 4.2.2.C.1), 21
    Long- and Short-term Planning (Adder 4.2.3.C.2), 23

## 162 Index

Management Commitment (Adder 4.21.2.C.1), 47
Migration Planning (Adder 4.4.2.S.4), 67
New Product Introduction (NPI) (Adder 4.21.4.C.1), 48
Notification About Problems (Adder 4.19.C.3), 43
Operational Changes (Adder 4.9.HV.1), 56
Operator Qualifications (Adder 4.9.HV.2), 56
Packaging and Labeling Audit (Adder 4.10.4.H.2), 62–63
Patch Documentation (Adder 4.15.6.S.1), 72
Patching Procedures (Adder 4.19.S.1), 72
Periodic Retesting (Adder 4.4.8.H.1), 58–59
Problem Escalation (Adder 4.19.C.5), 44
Problem Resolution (Adder 4.19.S.2), 73
Problem Resolution Configuration Management (Adder 4.19.HS.2), 54
Problem Severity (Adder 4.19.C.4), 43
Process Measurement (Adder 4.20.1.C.1), 44
Product Identification (Adder 4.8.HS.2), 53–54
Project Plan (Adder 4.4.2.C.1), 27
Purchasing Procedure(s) (Adder 4.6.1.C.1), 35
Quality Improvement Concepts (Adder 4.18.C.2), 40
Quality Improvement Program (Adder 4.21.1.C.1), 45
Quality Objectives (Adder 4.1.1.C.1), 21
Release Management (Adder 4.9.S.2), 70
Replication (Adder 4.9.S.1), 69
Requirements Allocation (Adder 4.4.4.S.2), 68
Requirements Traceability (Adder 4.4.1.C.1), 27
Service Delivery Plan (Adder 4.9.V.2), 75
Service Performance Data (Adder 4.21.3.V.1), 76
Service Resources (Adder 4.19.C.2), 43
Services Design Output (Adder 4.4.5.V.1), 74
Software Used in Service Delivery (Adder 4.9.V.1), 75
Software Virus Protection (Adder 4.15.2.S.1), 71
Subcontractor Input (Adder 4.2.3.C.3), 24
Supplier Performance Feedback (Adder 4.21.1.C.3), 46
Supplier's Recall Process (Adder 4.19.H.1), 64
Supplier's Support Program (Adder 4.19.C.1), 42
Support Software and Tools Management (Adder 4.2.2.S.1), 65
Test Documentation (Adder 4.10.1.S.1), 71
Test Planning (Adder 4.4.2.C.2), 29
Testing of Repair and Return Products (Adder 4.10.4.H.1), 62
Tool Changes (Adder 4.4.9.V.1), 74
Traceability for Recall (Adder 4.8.H.1), 61
Traceability of Design Changes (Adder 4.8.H.2), 61
Tracking Changes (Adder 4.4.9.H.1), 60
Training Content (Adder 4.18.C.6), 41–42
Training Requirements and Awareness (Adder 4.18.C.3), 40
Trend Analysis (Adder 4.13.2.C.1), 36
Work Areas (Adder 4.15.1.C.1), 38
adders, 5
  as correspond to ANSI/ISO/ASQC Q9001–1994, 52
  common, 20
  definition of, 1
  list of, 52
  sources of, 7
  hardware, 58
  hardware/services, 55
  hardware/software, 52–53
  service, 73
  software, 65
Advanced Quality Training (Adder 4.18.C.5), 41
American Society for Quality (ASQ)
  additional responsibilities of, 118
  and BS 7799 certification, 110
  as Forum Administrator, 9, 15
  as RRS manager, 108
  interaction with metrics data flow, 111
  responsibilities of, 15
  role in RRS, 108
  *See also* Forum Administrator
ANSI/ISO/ASQ Q9000-3–1997
  as guide, 2
  *See also* ISO 9000-3
ANSI/ISO/ASQC Q9001–1994, 129
  as basis for requirements, 19
  as correspond to adders, 52
  as foundation of TL 9000, 5
  key additions to, 8
  requirements based on, 51

## Index 163

role in handbooks, 7
year 2000 revision of, 16
*See also* ISO 9001
ANSI/ISO/ASQ 9001:2000, 126
  consideration of, 19
  updating QMS to, 132
ANSI/ISO/ASQC Q9002–1994, 129
ANSI/ISO/ASQC Q9004–1994, year 2000 revision of, 16
Anti-Static Protection (Adder 4.15.1.C.2), 38
ASQ. *See* American Society for Quality (ASQ)
auditor, training course for, 14
Automotive Industry Action Group (AIAG), 2

## B

British Standards Institute (BSI), 10, 110
BS 7799, 10, 110
Business Excellence Acceleration Model (BEAM), 144

## C

change management, 60, 61
Change Management Process (Adder 4.4.9.C.1), 33
common metrics, 87
Component Changes (Adder 4.4.9.H.2), 60
Computer Resources (Adder 4.4.2.S.2), 66
confidentiality of source and data, 108
Configuration Management (CM) Plan (Adder 4.8.HS.1), 53
Content of Requirements (Adder 4.4.4.H.1), 58
Content of Testing (Adder 4.4.8.H.2), 59
Contract Review (notes of section 4.3), 25–26
Control of Customer-Supplied Documents and Data (Adder 4.5.1.S.1), 69
Control of Inspection, Measuring, and Test Equipment (section 4.11), 63
Control of Nonconforming Product (section 4.13), 36
Corrective and Preventive Action (notes of section 4.14), 37
corrective patch quality (CPQ), 100
cost of quality, 77
Course Development (Adder 4.18.C.1), 39
Customer and Subcontractor Input (Adder 4.4.4.C.1), 31

Customer Involvement (Adder 4.2.3.C.1), 22
Customer Requested Date (CRD), 90
Customer Satisfaction (Adder 4.21.3.C.1), as linked to customer involvement, 23
Customer Satisfaction (Adder 4.21.3.C.1), 48
Customer Supplier Quality Program (CSQP)$^{SM}$**, 129, 132
customer-supplier communication
  additional requirements for, 8
  and use of metrics, 120
  during conformance, 137
Customer-Supplier Communication (Adder 4.21.2.C.2), as linked to customer involvement, 23
Customer-Supplier Communication (Adder 4.21.2.C.2), 47
customer-supplier quality improvement program
  benefits of, 23
  establishment of, 22
customer-supplier relationship
  and common metrics, 87
  and use of metrics, 86
  as linked to metrics, 78
  importance of, 2
  management's role in, 47

## D

data
  confidentiality of, 108
  integrity, and RRS/MRS, 109
  security, elements of, 108
  submission, training course for, 14
  transmission failure, resolution of, 116
Design Control (section 4.4)
  common adders, 26
  hardware adders, 58
  service adders, 74
  software adders, 66
Design Output (Adder 4.4.5.S.1), 68
Design Requirements (Adder 4.4.4.C.2), as linked to requirements allocation, 68
Design Requirements (Adder 4.4.4.C.2), 31
Design Validation (note of section 4.4.8), 55
Design Validation (section 4.4.8), 33
Design Verification (section 4.4.7), 33
Deterioration (Adder 4.15.3.H.1), 64
Disaster Recovery (Adder 4.2.3.C.4), 24

Document and Data Control (section
        4.5), 69
double blind system, 108
    establishment of, 9
    purpose of, 111

# E

EIRUS relationship with QuEST
        Forum, 124
electrostatic discharge. *See* ESD
Emergency Service (Adder 4.19.HS.1), 54
Employee Participation (Adder
        4.21.1.C.2), 46
Employee Skills List (Adder 4.9.HV.3), 56
encryption process, 109
End of Life Planning (Adder 4.4.2.C.3), 30
ESD, damage by, 38–39
ESD Training (Adder 4.18.C.4), 41
ESD-resistive products, handling of, 41
Estimation (Adder 4.4.2.S.1), 66
European In-Process Quality Metrics
        (E-IPQM), 124
    and RQMS User Group, Forum
        relationship with, 16
European Reliability and Quality
        Measurements for Telecommunica-
        tions Systems (E-RQMS), 124
Excel Partnership
    additional training course, 14
    as course provider, 8
    role in training, 13, 123
external support, as concern during
        TL 9000 registration, 132

# F

feature patch quality (FPQ)
    definition of, 100
    metric, as based on RQMS, 100
Field Performance Data (Adder
        4.21.3.H.1), 64–65
first year return, rate of, 94
fix response metric, 89
fix response time (FRT), definition of, 88
Forum. *See* QuEST Forum
Forum Administrator
    and use of metrics, 86
    role in double blind system, 9
    *See also* American Society for Quality
Frequency of Testing (Adder 4.4.8.H.3), 59

# G

gap analysis
    benefits of, 136
    during implementation process, 135–136
goal-question-metric (GQM), 81
    sample questions, 82
GR-1202, role in handbooks, 7
GR-1252, role in handbooks, 7
GR-929-CORE, 10, 122

# H

Handling, Storage, Packaging, Preservation,
        and Delivery (section 4.15)
    common adders, 38
    hardware adder, 63
    software adders, 71
hardware only metric, application of, 94

# I

Identification of Software Requirements
        (Adder 4.4.4.S.1), 68
Identified Equipment (Adder
        4.11.2.H.1), 63
IEC/TC 56, QuEST Forum status with, 16
implementation
    costs as concern during TL 9000
        registration, 131
    management of process, 135
    training course for, 14
indicators
    definition of, 10
    use of, 122
Informing Customers (Adder 4.4.9.C.2), 34
Inspection and Test Documentation (Adder
        4.10.1.HV.1), 57
Inspection and Test Records (Adder
        4.10.5.HV.1), 57
Inspection and Testing (Adder 4.9.H.1), 62
Inspection and Testing (section 4.10)
    hardware adders, 62
    hardware/service adders, 57
    note of common adder, 36
    software adder, 71
Installation Plan (Adder 4.19.HS.3), 55
installation services, requirements of, 32
Integration Planning (Adder 4.4.2.S.3), 67
internal auditing, during
        implementation, 138
internal audits, benefits of, 136

# Index

ISO 9000
    and use of metrics, 120
    registration cost of, 131
ISO 9000-3, role in handbooks, 7
ISO 9001, 2
    role in handbooks, 6
    role in registration process, 13
ISO 9002, role in registration process, 13
ISO 9004-2, role in handbooks, 7
ISO 10011-1, role in handbooks, 7
ISO 10011-2, role in handbooks, 7
ISO 10011-3, role in handbooks, 7
ISO 14001
    registration cost of, 131
    TL 9000 integration with, 130
ISO/IEC 12207, 2
    role in handbooks, 7
ISO/IEC JTC1/SC7, Forum status with, 16
ISO/TC 176
    as basis for project planning techniques, 126
    Forum status with, 16

# L

Lifecycle (Adder 4.2.2.C.1), 21
lifecycle model (LCM), 139
Long- and Short-term Planning (Adder 4.2.3.C.2), 23
long-term return (LTR), rate of, 94

# M

management commitment, 130
Management Commitment (Adder 4.21.2.C.1), 47
Management Responsibility (section 4.1), 21
metric
    hardware only, 94
    problem report fix, 89
    RAA, 98
    service only, 102
    software only, 96
    system outage, 93
metrics
    benefits of, 78
    categories of, 6
    computation of, 121
    cost and performance-based, 1
    data storage, description of, 110
    during implementation process, 135
    goals of, 81
    ground rules, 79
    identifiers for OTD, 90
    importance of, 77, 134
    interactions within implementation process, *112*
    list of, 10
    model of, *85*
    obstacles in use of, 139
    package criteria of, 80
    Phase 1, 84
    product identification for, 122
    purpose of submission, 78
    selection using the GQM model, 81
    submission and analysis of, 105
    use of, 86
    use of, to improve product and service quality, 8–9, 120
    use in case studies, 139
    work group, 79
metrics data flow
    interaction of, 111
    interactions among suppliers, ASQ, and UTD, 116, 117
    interactions among suppliers, registrar, and ASQ, 117, 118
    interactions between ASQ and UTD, 118, 119
    interactions between supplier and ASQ, 115
    interactions between supplier and customers, 114
    interactions between supplier and UTD, 115, 116
    interactions during, 112, 113
Metrics Handbook, purpose of, 77
Metrics Repository System (MRS)
    and BS 7799 compliance, 110
    and use of metrics, 86
    and UTD, 107
    requirements of, 107–108
    role in double blind system, 9
    UTD's responsibilities of, 15
    versus RRS, 108
Migration Planning (Adder 4.4.2.S.4), 67
Mroz, Jim, 125
MRS. *See* Metrics Repository System (MRS)

# N

Network Equipment-Building System, 11
New Product Introduction (NPI) (Adder 4.21.4.C.1), 48

no trouble found (NTF), 65
normalization factors, identification of, 10, 121
Notification About Problems (Adder 4.19.C.3), 43
number of problem reports (NPR), 87

## O

On-Time Delivery (OTD), 90
On-Time Installed System Delivery (OTIS), and OTD, 90
On-Time Items Delivery (OTI), and OTD, 90
On-Time Service Delivery (OTS), and OTD, 90
Operational Changes (Adder 4.9.HV.1), 56
Operator Qualifications (Adder 4.9.HV.2), 56
Organizational and Technical Interfaces (section 4.4.3), 31
overdue problem report and fix responsiveness (OFR), definition of, 88

## P

Packaging and Labeling Audit (Adder 4.10.4.H.2), 62–63
Patch, versus software update, 96–97
Patch Documentation (Adder 4.15.6.S.1), 72
Patching Procedures (Adder 4.19.S.1), 72
Periodic Retesting (Adder 4.4.8.H.1), 58–59
pilot organizations, conformance problems of, 136
pilot program
 and QMS conformance, 138
 feedback goals of, 125
 goals of, 12
 launch of, 125
 participation in, 12
 participants in, 126
Plan–Do–Check–Act (PDCA), use of, 24
pre-assessment audit, benefits of, 140
Problem Escalation (Adder 4.19.C.5), 44
problem report fix metric, 89
Problem Resolution (Adder 4.19.S.2), 73
Problem Resolution Configuration Management (Adder 4.19.HS.2), 54
Problem Severity (Adder 4.19.C.4), 43
Process Control (section 4.9)
 hardware adders, 62
 hardware/service adders, 55
 service adders, 75
 software adders, 69
Process Measurement (Adder 4.20.1.C.1), 44
product and service
 categories, 11, 122
 expectations for, 120
 quality, and use of metrics, 120
product category tables, 6, 80, 122
Product Delivery (section 4.8), 53
Product Identification (Adder 4.8.HS.2), as linked to servicing, 64
Product Identification (Adder 4.8.HS.2), 53–54
Product Identification and Traceability (section 4.8), 61
product recall, 54
Project Plan (Adder 4.4.2.C.1), 27
Purchasing (section 4.6), as linked to lifecycle, 22
Purchasing (section 4.6), 35
Purchasing Procedure(s) (Adder 4.6.1.C.1), 35

## Q

QMS
 conformance to, in pilot program, 138
 coverage of TL 9000 documents, 129
 during implementation process, 135
 revision during implementation process, 136
 update of, during implementation, 141
 updating of, to ANSI/ISO/ASQ Q9001–2000, 132
QS-9000, 2
 registration cost of, 131
 role in handbooks, 7
 TL 9000 integration with, 130
Quality Improvement and Customer Satisfaction (section 4.21)
 as additional TL 9000 element, 5
 common adders, 45
 hardware adder, 64
 service adder, 76
Quality Improvement Concepts (Adder 4.18.C.2), 40
Quality Improvement Program (Adder 4.21.1.C.1), 45
Quality Objectives (Adder 4.1.1.C.1), 21
Quality Planning (section 4.2.3), 22
Quality System (section 4.2), 21, 65
QuEST Forum
 administration of, 14–15

# Index  167

creation of, 1
goal of, 1
global approach of, 124
member products and services,
    expectations for, 120
members
    benefits for, 4–5
    expectations of, 3
    results for, 5
membership
    categories, 4
    industry benefits of, 119
    new initiatives of, 143–144
    relationship with EIRUS, 124
    web site, 15

# R

registrar
    costs as concern during TL 9000
        registration, 131
    interaction with metrics data flow, 111
    role in metrics compliance, 11
    training courses for, 14
Registrar Accreditation Board (RAB), role
    in registration process, 13
registration, process of, 13
registration repository system (RRS)
    ASQ's role in, 108
    management of, 108
    MRS, and data integrity, 109
release application abort (RAA), 98
Release Management (Adder 4.9.S.2), 70
Reliability and Quality Measurements
    for Telecommunications
    Systems (RQMS)
    alignment of, during conformance, 137
    as basis for CPQ metric, 100
    as basis for FPQ metric, 100
    as basis for FRT, 88
    as basis for NPR, 87
    as basis for OFR, 88
    as basis for RAA metric, 98
    requirement of, 10
    versus SWU tracking, 97
    versus TL 9000 metrics, 121–122
Replication (Adder 4.9.S.1), 69
requirements
    codes for, 20
    codes used for identification, 51–52
    documents needed, 32
Requirements Allocation (Adder
    4.4.4.S.2), 68

Requirements Handbook, development of, 6
Requirements Traceability (Adder
    4.4.1.C.1), 27
return rate, 94
root cause analysis
    and problem resolution, 73
    facilitation of, 54

# S

sanctioned training courses, 123–124
    list of, 14
Service Delivery Plan (Adder 4.9.V.2), 75
service only metric, application of, 102
Service Performance Data (Adder
    4.21.3.V.1), 76
service quality, application of, 102
service quality metric
    examples of, 103 thru 104
    purpose of, 102
Service Resources (Adder 4.19.C.2), 43
Services Design Output (Adder
    4.4.5.V.1), 74
Servicing (section 4.19)
    common adders, 42
    hardware adder, 64
    hardware/software adders, 54
    software adders, 72
*shall* vs. *should*, 19, 51
software only metrics, application of, 96
software update (SWU) metric, 97
software update quality (SWU), 96 thru 97
Software Used in Service Delivery (Adder
    4.9.V.1), 75
Software Virus Protection (Adder
    4.15.2.S.1), 71
Standards Council of Canada, role in
    registration process, 13
STAT-A-MATRIX (The SAM Group)
    as course provider, 8
    role in training, 13, 123
Statistical Techniques (section 4.2), 44
Subcontractor Input (Adder 4.2.3.C.3), 24
Supplier Performance Feedback (Adder
    4.21.1.C.3), 46
Supplier's Recall Process (Adder
    4.19.H.1), 64
Supplier's Support Program (Adder
    4.19.C.1), 42
Support Software and Tools Management
    (Adder 4.2.2.S.1), 65
system outage, 92
system outage metric, computation of, 93

# 168  Index

## T

Telcordia Technologies, 10, 16, 121–122
  as basis for metrics, 124
  as linked to informing customers, 34
  documents of, 7
  use of requirements in training, 132
Telecommunication Quality Index,
    establishment of, 87
telecommunications industry needs, 2
Test Documentation (Adder 4.10.1.S.1) 71
Test Planning (Adder 4.4.2.C.2), 29
Testing of Repair and Return Products
    (Adder 4.10.4.H.1), 62
TickIT, 129
TL 9000
  benefits of, 3
  goals of, 3
  Handbooks, 1, 6
  implementation
    aspects of, 138
    recommendations to
        organizations, 141
  metrics, goals and expectations of, 118
  Model, 6
  Quality System Metrics Handbook,
      publication and distribution of, 15
  Quality System Requirements Handbook,
      publication and distribution of, 15
  registration, costs of, 131–133
  Requirements, during implementation
      process, 135
Tool Changes (Adder 4.4.9.V.1), 74
total effort, as concern during TL 9000
    registration, 133
total staff time, as concern during TL 9000
    registration, 133
TR-179, 2
Traceability for Recall (Adder 4.8.H.1), as
    linked to servicing, 64
Traceability for Recall (Adder 4.8.H.1), 61
Traceability of Design Changes (Adder
    4.8.H.2), 61
Tracking Changes (Adder 4.4.9.H.1), as
    linked to traceability, 61
Tracking Changes (Adder 4.4.9.H.1), 60

training, 13
  as concern during TL 9000
      registration, 132
  description of sanctioned courses, 14
  during implementation, 138
  Excel Partnership's role in, 13, 123
  intellectual property rights of, 13
  STAT-A-MATRIX (The SAM Group)
      role in, 13, 123
Training (section 4.18), 39
Training Content (Adder 4.18.C.6), 41–42
Training Requirements and Awareness
    (Adder 4.18.C.3), 40
Trend Analysis (Adder 4.13.2.C.1), 36
TR-NTW-000179, role in handbooks, 7

## U

University of Texas at Dallas (UTD)
  and BS 7799 compliance, 10
  and development of double blind
      system, 108
  and MRS, 107
  as administrator of MRS, 15
  interaction with metrics data flow, 111
  metric reported to, 88
  MRS responsibilities of, 15
  responsibilities of, 120
  retention of, 9
  role in double blind system, 9
UTD. *See* University of Texas at
    Dallas (UTD)

## W

Work Areas (Adder 4.15.1.C.1), 38
work groups
  and metric alignment, 16
  establishment of, 1
  member interaction with, 5
  metric, 79
  purpose of, 1
  use of project planning techniques, 12